HANDBOOK OF ELECTRONIC MATERIALS
Volume 6

HANDBOOK OF ELECTRONIC MATERIALS

Compiled by:
ELECTRONIC PROPERTIES INFORMATION CENTER
Hughes Aircraft Company
Culver City, California

Sponsored by:
AIR FORCE MATERIALS LABORATORY
Air Force Systems Command
Wright-Patterson Air Force Base, Ohio

HANDBOOK OF ELECTRONIC MATERIALS
Volume 6

Silicon Nitride for Microelectronic Applications

Part 2
Applications and Devices

John T. Milek

Electronic Properties Information Center
Hughes Aircraft Company, Culver City, California

IFI/PLENUM · NEW YORK-WASHINGTON-LONDON · 1972

This document has been approved for public release and sale;
its distribution is unlimited. Sponsored by: Air Force Materials
Laboratory, Wright-Patterson Air Force Base, Ohio.

Library of Congress Catalog Card Number 76-147312

©1972   IFI/Plenum Data Corporation, a Subsidiary of
Softcover reprint of the hardcover 1st edition 1972
Plenum Publishing Corporation
227 West 17th Street, New York, N.Y. 10011

United Kingdom edition published by Plenum Press, London
A Division of Plenum Publishing Company, Ltd.
Davis House (4th Floor), 8 Scrubs Lane, Harlesden, NW10 6SE, London, England

ISBN-13: 978-1-4615-9611-0   e-ISBN-13: 978-1-4615-9609-7
DOI: 10.1007/ 978-1-4615-9609-7

FOREWORD

This survey is concerned with the use of silicon nitride in the semi-conductor and microelectronics industries. The Handbook of Electronic Materials, volume 3, comprises part 1 of this survey and includes preparation and properties information.

This report was prepared by Hughes Aircraft Company, Culver City, California under Contract Number F33615-70-C-1348. The work was administered under the direction of the Air Force Materials Laboratory, Air Force Systems Command, Wright-Patterson Air Force Base, Ohio, with Mr. B. Emrich, Project Engineer.

The Electronic Properties Information Center (EPIC) is a designated Information Analysis Center of the Department of Defense, authorized to provide information to the entire DoD community. The purpose of the Center is to provide a highly competent source of information and data on the electronic, optical and magnetic properties of materials of value to the Department of Defense. Its major function is to evaluate, compile and publish the experimental data from the world's unclassified literature concerned with the properties of materials. All materials relevant to the field of electronics are within the scope of EPIC: insulators, semiconductors, metals, super-conductors, ferrites, ferroelectrics, ferromagnetics, electroluminescents, thermionic emitters and optical materials. The Center's scope includes information on over 100 basic properties of materials; information generally regarded as being in the area of devices and/or circuitry is excluded.

CONTENTS

INTRODUCTION

Bulk silicon nitride has found a number of applications in the refractories field because of its high melting point, hardness, low thermal expansion, and good thermal shock resistance. It is chemically inert and this fact, together with the previous qualifications, makes it useful for crucibles, rods, discs and boats. However, no electronic applications seem apparent for the bulk form despite its good electrical properties; the resistivity is 10^{12} ohm-cm at room temperature and 10^6 ohm-cm at 1000°C.

With the marked advances in silicon integrated circuit technology and the search for greater reliability, serious consideration has been given to silicon nitride films. A cursory analysis of the semiconductor literature in the past few years has, therefore, indicated a constantly increasing interest in preparing and using silicon nitride as an insulating, masking, or passivating layer. Such silicon nitride layers can be prepared by a variety of deposition processes as detailed in the companion volume, HANDBOOK OF ELECTRONIC MATERIALS, VOL. 3, Silicon Nitride for Microelectronic Applications: Part 1, Preparation and Properties, and is summarized in Figure 1.

The growing interest in silicon nitride is reflected in Table 1 which lists the patents issued on this material in recent years.

In view of its improved electrical properties and superior chemical resistance (see HANDBOOK OF ELECTRONIC MATERIALS, VOL. 3), it was natural for silicon nitride to attract attention as a substitute for silicon dioxide insulating layers in integrated circuits and discrete devices; silicon nitride resists contamination much more effectively than SiO_2, and reduces threshold voltage. The advantages of thin film silicon nitride (compared with silicon dioxide, normally used for these purposes) are, first of all, the higher dielectric constant and imperviousness to the diffusion of impurities. Although several years ago, silicon nitride was a very poorly understood material, it has

1

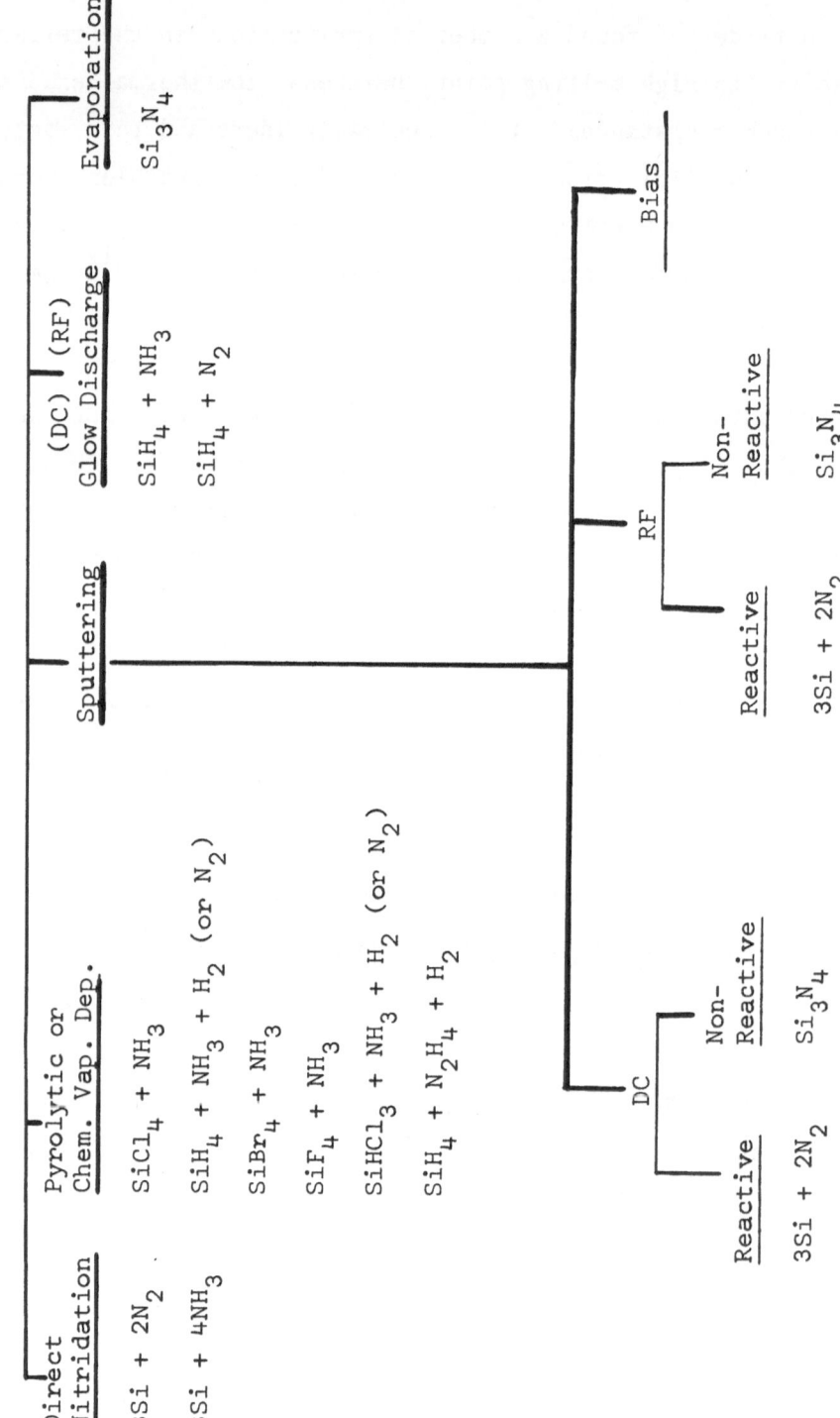

Figure 1. Silicon Nitride Deposition Techniques.

Table 1. Patents Issued Concerning Thin Films Of Silicon Nitride

		United States	
Number	Date	Inventor	Subject
3,565,674	28 Feb 1971	B.W. Boland et al	Deposition
3.560,364	2 Feb 1971	P.J. Burkhardt	Unsupported Films
3,558,348	26 Jan 1971	M.J. Rand	Films for Semiconducting Devices
3,427,506	1 July 1969	T. Okumura	FET With Insulated Gates
3,427,514	11 Feb 1969	J. Olmstead et al	MOS Tetrode
3,421,936	14 Jan 1969	F.L. Vogel Jr.	Coating On Semiconducting Bodies
3.401,450	17 Sept 1968	G.C. Godejahn, Jr.	Isolation
3,385,729	28 May 1968	G.A. Larchian	Isolation
3,373,051	12 Mar 1968	T.L. Chu et al	Vapor Deposition
3,372,067	5 Mar 1968	H. Schäfer	Vapor Deposition
3,320,484	16 May 1967	J.E. Riley and B.B. Williams	Dielectric Devices
3,287,243	22 Nov 1966	J.R. Ligenza	Glow Discharge, Sputtering
3,246,214	12 Apr 1966	F.B. Hugle	Differential Masking
3,226,194	28 Dec 1965	U.E. Kuntz	Coating
3,165,430	12 Jan 1965	F.B. Hugle	Masking
3,149,398	22 Sept 1964	J.L. Sprague et al	Capacitor Dielectric
3,122,450	25 Feb 1964	C.R. Baines et al	Deposition
3,095,527	25 June 1963	C.R. Baines et al	Capacitor Dielectric
3,030,243	12 June 1962	C.R. Baines et al	Dielectric
		France	
1,492,719	18 Aug 1967	V.Y. Doo et al	Masking
		South Africa	
6,800,059	28 Aug 1968	H.F. Sterling et al	Deposition

Table 1. (Continued)

England

Number	Date	Inventor	Subject
1,106,510	20 Mar 1968	H.F. Sterling et al	Deposition
1,118,757	3 July 1968	W.N. Pennebaker	Deposition
1,133,333	13 Nov 1968	D.R. Ashworth et al	Insulating Coating
1,134,964	27 Nov 1968	A.G. Siemens	Deposition
1,134,352	20 Nov 1968	A.G. Siemens	Deposition
1,130,138	9 Oct 1968	N.C. Tombs	Dielectric

France

1,492,719	18 Aug 1967	V.Y. Doo et al	Masking

East Germany

60,197	5 Feb 1968	H.J. Schnabel and F. Fleischer	Deposition

Netherlands

6,606,405	13 Nov 1967	N.V. Philips Gloeilampenfab.	Deposition
6,707,515	2 Dec 1968	N.V. Philips Gloeilampenfab.	Deposition
6,708,533	27 Dec. 1967	N.V. Philips Gloeilampenfab.	Deposition

since been investigated for a wide range of applications. However, the silicon-silicon nitride interface is known to have a relatively high density of surface states and studies have tried to overcome this shortcoming. The result has been the exploration of silicon oxynitride, silicon nitride-silicon oxide MNOS structures, silicon dioxide-silicon nitride-silicon dioxide or MONOS structures. Further, consideration has been given to aluminum oxide and other oxide layers. The following diagram illustrates the trends occurring in metal-insulation-semiconductor (MIS) devices or structures in an attempt to find a superior dielectric insulator:

Figure 2. Metal-Insulation-Semiconductor Devices

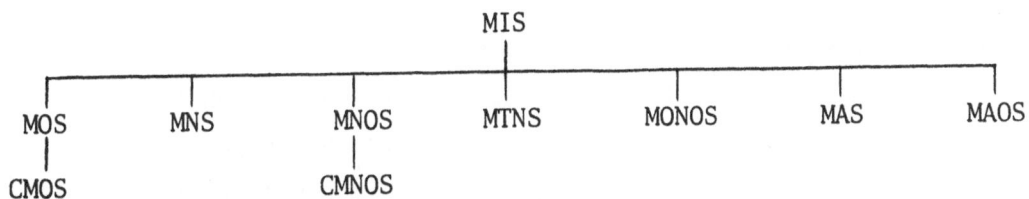

CMNOS	Complementary - Metal - Nitride - Oxide - Silicon (or semiconductor)
CMOS	Complementary - Metal - Oxide - Silicon (or semiconductor)
MAOS	Metal - Aluminum Oxide - Silicon Oxide - Silicon
MAS	Metal - Aluminum Oxide - Silicon (or semiconductor)
MIS	Metal - Insulation - Semiconductor
MNS	Metal - Nitride - Silicon (or semiconductor)
MNOS	Metal - Nitride - Oxide - Silicon (or semiconductor)
MONOS	Metal - Oxide - Nitride - Oxide - Silicon (or semiconductor)
MOS	Metal - Oxide - Silicon (or semiconductor)
MTNS	Metal - Thick Silicon Nitride - Silicon (or semiconductor)

It is evident that a wide range of MIS structures have been evaluated or explored to date: silicon dioxide layers, silicon nitride layers, silicon oxynitride, silicon dioxide-silicon dioxide composite layers, thick film silicon dioxide-silicon nitride-silicon dioxide layers, aluminum oxide layers, aluminum oxide-silicon dioxide layers. From a functional or device application perspective, it is also possible to discern other patterns of evolution. To be sure, each dielectric has a certain combination of properties and characteristics which may or may not make it suitable for every MIS device application or function. There has always been competition between materials: cost, processing ease, handling, packaging, advantages, limitations, etc. The results of this study show that silicon nitride films play an important role in modern solid state technology and may one day become the workhorse of the microelectronics field, a position held for the past ten years by silicon dioxide.

In the development of metal-insulator-silicon (MIS) integrated circuits, there is still need for a suitable thin film insulating layer which can be prepared by a single-stage process. As compared with the metal-nitride-oxide-silicon (MNOS) system currently in use, a nitride layer would bring the advantages of greater simplicity and economy to several stages in the preparation of such devices and circuits. Thus, the following chapters will present an analysis of the various structures or devices which have made use of nitride layers as well as combinations of nitride-oxide with the resulting electrical characteristics, advantages and disadvantages, and inherent problems of the structures.

There are several potential advantages in using silicon nitride alone for the fabrication of silicon devices. It is a very effective barrier to the diffusion of impurity ions and has a dielectric constant of approximately 8 (compared to approximately 3.9 for SiO_2). In addition, it is hard and abrasion resistant. Unfortunately, as shown in another section of this report, the amorphous silicon nitride films grown by chemical vapor deposition (CVD) or pyrolytic deposition form an Si_3N_4/Si interface which develops a high, and time-variable, charge density under conditions of normal device operation.

In general, the nitride insulating layer (a) must be a satisfactory diffusion mask (in particular, charged impurities such as Na^+ ions must show negligible drift in the electric field gradients imposed during device operation) and (b) it must form an interface with the silicon substrate which meets certain specifications with regard to surface states and surface charge density (for many applications, the relatively low value of 10^{11} charges per cm^2 is required together with an absence of hysteresis effects when the device is used as an "on-off" switch).

Depending on the purpose of the device, interface charge can be a boon or a bane in metal-nitride-oxide semiconductors. In a digital integrated circuit made by the MNOS process, for example, the charge is highly undesirable because it makes the threshold voltage unstable. In an MNOS memory, on the other hand, interface charge is the mechanism by which data is stored. In part 1 are described the various ways that the charge can be controlled by varying the composition of the silicon nitride layer. For successful device fabrication and application purposes, it is important to be aware of the established relationships among interface charge, electronic conductance, and deposition conditions, as well as charge and discharge times as a function of voltage and temperature.

Metal-nitride-silicon (MNS) and metal-nitride-oxide-silicon (MNOS) structures have been the subject of extensive investigations in recent years. Tombs et al. * had proposed the MNS structure as a substitute for MOS (metal-oxide-silicon) transistors, claiming improvements in device stability, dielectric strength of the gate, and control of surface-state density at the dielectric-silicon interface. Subsequent investigations by Chu and coworkers, Hu and coworkers, and Pao and O'Connell reported the presence of charge instabilities in the MNS structure, resulting in hysteresis behavior of the charge accumulated at the

* References listed alphabetically by author in the bibliography.

silicon nitride-silicon interface as a function of applied gate voltage. To avoid this unstable behavior, Chu et al. (1)** suggested that a thermal oxide layer be inserted between the silicon nitride layer and the silicon substrate, resulting in a stable MNOS field-effect transistor, which combines the low surface-state density of the thermal oxide-silicon interface with the passivation properties of the silicon nitride layer.

On the other hand, the same structure with a thin thermal oxide layer (<500 Å) has recently been shown by Wallmark and Scott Jr. to exhibit hysteresis behavior of turn-on voltage as a function of applied gate voltage, resulting from stable charge storage at the silicon nitride-silicon dioxide interface. The storage function associated with the hysteresis characteristic, leads to the potential application of variable turn-on voltage MNOS transistors in random-access and alterable read-only semiconductor memories, as well as digital circuits with nonvolatile storage capability. Further details on these devices and applications will be discussed in the various sections of this book.

Naber has reported on a method of reducing the mobile electric charge in MNOS structures which involves heating in hydrogen from 1 to 4 minutes at 1200°C. The shift in flatband voltage for other temperature levels and times are shown in Table 2. He states, however, that the mechanism by which the quantity of mobile charges is reduced by the high-temperature hydrogen treatment is not fully understood.

Figure 3 arbitrarily classifies the various possibilities as well as applications, as reported in the literature.

** Number in parenthesis indicates specific reference published by the author.

Table 2. Flat-band Voltage Shifts of Hydrogen
Processes MNOS Structures. (Naber)

Temp. °C	Time, min.	ΔV_{FB},V
1000	1	48
	2	109
	2	100
	4	17
	4	86
1100	1 .	46
	2	44
	4	8
1150	2	5.0
	2	2.0
1200	1	10
	2	2.0
	4	2.0

Silicon Nitride Applications

Diffusion Mask	Gate Dielectric	Passivation Purposes	Logic	Capacitors	Diodes and Rectifiers	Memory	Switching Elements
Si	MNST	Si	MNOS	MIS	MNS	MNOS	MIM
Ge	MNOST	Ge		MIM		MNS	
GaAs	MNSFET	GaAs		MNOS		MTNS	
	MNOSFET						
	Silicon						
	GaAs						

MIM   Metal – Insulator – Metal
MIS   Metal – Insulation – Semiconductor
MNOS   Metal – Nitride – Oxide – Silicon (or semiconductor)
MNOS FET Metal – Nitride – Oxide – Silicon – Field Effect Transistor
MNOST Metal – Nitride – Oxide – Semiconductor Transistor
MNS   Metal – Nitride – Silicon (or semiconductor)
MNSFET Metal – Nitride – Silicon Field Effect Transistor
MNST   Metal – Nitride – Silicon Transistor
MTNS   Metal – Thick Silicon Nitride – Silicon (or semiconductor)

Figure 3. Classification of Silicon Nitride Applications.

DIFFUSION MASK APPLICATIONS

The excellent chemical stability of silicon nitride makes its use attractive in the area of device passivation and diffusion masking. As a result, it has been widely studied by many investigators for use with semiconductor substrates: silicon, germanium, and gallium arsenide, in planar device technology. Films suitable for diffusion mask applications must, furthermore, remain continuous and impervious at temperatures up to 1100°C. This creates a problem for those materials which are deposited at lower temperatures (e.g. 300 to 600°C). Lower temperature deposited materials will usually crack or leave circular holes. Silicon nitride is a considerably denser material than SiO_2 (3.18 vs 2.2 g/cm^3, respectively) and does not readily permit migration or diffusion through it.

As early as 1966, it was reported that the Components Division of IBM and the Sperry Rand Research Center were investigating silicon nitride for use as a diffusion mask as a replacement for silicon dioxide. Silicon nitride was deposited at relatively low temperature (approximately 800°C) while SiO_2 requires 950°C. In addition to being impervious to diffusion by boron, phosphorus, and arsenic, silicon nitride made it possible to use other important diffusants such as gallium, zinc, and oxygen; silicon dioxide, in contrast, is permeable to these materials.

Burgess and coworkers at Sprague Electric Company studied the ability of silicon nitride layers deposited on silicon oxide-passivated silicon to act as contamination barriers through the use of a radioisotope, sodium-22. They concluded that silicon nitride films deposited over silicon oxide layers may be used to shield the passivating oxide film against sodium penetration. These films, deposited by a variety of techniques, act not only as sodium barriers but also as sodium getters. The film deposition temperature strongly influences the sodium gettering characteristic of the silicon nitride layer. Films pyrolytically deposited at the lower temperature (approximately 850°C) are always better sodium getters than those produced at 1000°C or 1100°C, but all the chemical vapor deposited films give equal protection to the silicon oxide layer against sodium penetration. According to these investigators, this fact indicates that the films produced at 1000°C or 1100°C are better mechanical barriers than those prepared at 850°C. Further, the presence of moisture in an annealing atmosphere causes sodium to move into the silicon oxide-silicon interface from the silicon nitride layer. This effect is more pronounced in systems that contain the better sodium gettering silicon nitride films.

In a recent paper, Fränz and Langheinrich (1) found that sodium is gettered and enriched in the nitride portion of silicon dioxide-silicon nitride double layer films. Their silicon nitride films used in the experiments were deposited pyrolytically at 1000°C from silane and ammonia. Sodium-24 isotopes and radio-chemical methods were used to evaluate the sodium concentrations in the nitride films. Annealing experiments were also conducted from 600 to 1200°C for periods of ten minutes and the residual sodium contents vs. temperature of annealing measured. Various graphs were presented by these authors summarizing their experimental results on the distribution of sodium in the silicon nitride films.

The ability of silicon nitride to mask the diffusion of indium in silicon was demonstrated by Fränz and Langheinrich (2); a silicon dioxide film, five times as thick as the silicon nitride, failed to mask this dopant in diffusion experiments conducted at 1000°C for 10 minutes. They also studied (3) silicon nitrides as a mask in phosphorus pentoxide diffusion. Experimental results showed there were analogous glass formations of silicon nitride with P_2O_5 and SiO_2 with P_2O_5 and the masking effects therefore were similar.

Chu et al. (2) studied the masking ability of silicon nitride amorphous films against the diffusion of aluminum, boron and phosphorus into silicon. These dopants were found to react with silicon nitride in varying degrees at high temperatures, and these reactions must be considered in determining the thickness of silicon nitride film required for masking purposes. A film of 500 Å thickness deposited by the nitridation technique was found to be effective in masking the diffusion of boron under their deposition conditions; 1300 Å thickness was sufficient to mask the diffusion of phosphorus. In similar fashion,

12

Hu and Doo (1) studies on the diffusion masking of silicon nitride films on silicon have shown that very thin layers less than 1200 Å of nitride would mask against B, Ga, P, and As. They suggested that Si_3N_4 might even be used as a universal diffusion mask. Their shallow (< 9μ) junction diffusion studies indicated that no masking failures had been encountered. Hartman and Herr (2), conducted qualitative studies on the masking properties of Si_3N_4. In particular, a range of diffusion conditions and masking properties were studied for some of the common doping sources including B, P, Ga, and As. Their results indicated that S_3N_4 was not a diffusion mask for all conditions, particularly for the very thin films and for the higher diffusion temperatures. Heumann et al. in 1968 also pointed out that thin (< 1500 Å) nitride films will mask for a certain set of diffusion conditions, however, it should not be concluded that silicon nitride can be used as a universal diffusion mask. Various types of masking failures occur for temperatures and times required for deep junction diffusions. Table 3 summarizes their results for various dopants: B, P, Ga, and As. These investigators noted that the masking ability of the nitrides is extremely dependent on the chemical nature of the diffusion conditions. They concluded that generally speaking, thin silicon nitride films (1500 Å thick) can be used as diffusion masks for the times and temperatures suitable for shallow (< 9μ) junctions if the proper diffusant sources are used. Thin film nitride masking and much deeper silicon junction depth (30μ) can be attained by using higher temperatures if boron is the diffusant and a sealed tube system is used.

Gregor (1) has found that silicon nitride is an extremely effective barrier against atomic, molecular, or ionic diffusion. He found that the deposition of silicon nitride films by RF sputtering (reactive) is a feasible and practical method for producing passivation and diffusion mask films at moderate substrate temperatures. The physical and electrical properties of such films are determined by sputtering parameters such as nitrogen pressure, RF power, substrate temperature, and residual pressure of the vacuum chamber. By careful control of these factors, Si_3N_4 films with suitable etch rates for conventional photolithography can be produced. Other experiments showed that when good quality sputtered Si_3N_4 films were obtained, substances such as water, oxygen, phosphorus, and gallium did not diffuse through the films at temperatures of 1050 to 1200°C.

Table 3. Silicon Nitride Diffusion Masking. (Heumann et al.)

Dopant	Diffusion system	Thickness (Å)	Temp. (°C)	Time (hr)	Masked	Masked junct. depth. μ*	Unmasked junct. depth. μ	Comments
B	Sealed Tube	>300	1100	4	Yes	—	6	
B	Sealed Tube	>400	1200	20	Yes	—	25	
B	Sealed Tube	>500	1200	30	Yes	—	25	Film conversion
B	Sealed Tube	1000	1250	10	Yes	—	30	
B	Sealed Tube	<600	1250	10	No	Spikes	30	Spiking failure
B	Box	500-1500	1100	5	Yes	—	4	
B	Box	1500	1200	5	No	—		Film conversion
P	Sealed Tube	>1500	1100	3	Yes	—	—	
P	Sealed Tube	<1000	1100	3	No	—	—	
P	Sealed Tube	>1500	1100	5	Yes	—	5	
P	Sealed Tube	>1500	1100	10	Yes	—	9	
P	Sealed Tube	<1100	1100	10	No	—	9	
P	Sealed Tube	1500	1200	4	No	3	12	
P	Box	1000	1100	1	No	—	6	Film conversion
Ga	Sealed Tube	1000	1100	4	Yes	—	6	
Ga	Sealed Tube	300	1185	15	No	10	18	
Ga	Sealed Tube	1200	1185	15	Yes	Spikes	—	Localized spiking
Ga	Sealed Tube	1500	1200	4	Yes	Spikes	14	Localized spiking
As	Sealed Tube	1500	1100	4	Yes	—	0.6	
As	Sealed Tube	1000	1150	6	Yes	—	1.8	
As	Sealed Tube	1500	1150	6	Yes	Spikes	1.8	Films cracked
As	Sealed Tube	1500	1200	20	No	6	11	

*Where mask failure is noted but no masked junction depth is given, masking failure was determined by using a thermoelectric probe.

14

Sugawara has reported on his work in hydrogen chloride vapor etching of silicon surfaces through windows in SiO_2 and Si_3N_4 masks and the formation of facets in order to understand selective epitaxial growth on silicon substrates. For specimens with silicon nitride films, facets were developed directly below the film as well as on the periphery of the bottom of the etched area.

In the fabrication of semiconductor devices and integrated circuits, successive diffusions to form p-n junction structures in selected regions is one of the most important processes, according to Doo (1). To limit the diffusion to selected regions, it is necessary to use a masking material which is impermeable to the diffusants. In addition, the masking material must allow sharp definition by photoengraving. This requires a masking material that can be easily removed with small undercut. For a given masking material, the thinner the layer, the smaller the undercut. As early as 1966, Doo reviewed the literature on silicon dioxide as a diffusion masking material in semiconductor device fabrication and found that it fails to mask many important diffusants such as gallium, zinc, oxygen, etc. Doo and coworkers were some of the early investigators of silicon nitride as a diffusion masking material. Their nitride films were prepared by the silane/ammonia reaction with substrate temperatures ranging from 750 to 1100°C, and a wide range of thicknesses were evaluated with different diffusants as shown in Table 4. They concluded from their studies that silicon nitride is an effective diffusion mask, impermeable to diffusants which silicon dioxide fails to mask. The minimum film thickness required to mask most common dopants in silicon and germanium is almost an order of magnitude smaller than silicon dioxide.

Table 4. Diffusion Parameters. (Doo (1))

Sub-strate	Si_3N_4 Thick-ness (Å)	Dif-fusant	Diffusion Temp(°c)/Time(min)		Thickness Factor $x_j(\mu)C_o$ (atoms/cc)	
			Deposit	Drive-in	w/out mask	w/out mask
Si	1200	B	980/10	1200/30	2.07	7×10^{19}
Si	1200	P	1100/10	1100/20	1.8	1×10^{21}
Si	150	As		1200/120	1.44	1.4×10^{20}
Si	250	Ga		1100/90	3.2	4×10^{19}
Si	250	O		1150/20	0.5	—
Ge	250	Ga		800/120	0.92	1×10^{19}

μ = thickness

Thin silicon nitride layers have been shown to prevent or retard the annihilation of fast states in underlying thermally oxidized silicon during the post-aluminum annealing treatment by Deal et al. (1). On varying their process conditions they noted the following effects:

1. The masking or retardation of the fast state annihilation in $Si_3N_4/SiO_2/Si$ structures due to the silicon nitride layer is not dependent on silicon impurity type, concentration, or silicon orientation. Nitride layers thicker than 200 Å are effective in retarding fast state reduction.

2. The final density of fast states is a function of the carrier and cooling ambient during silicon nitride deposition where $SiCl_4$ is the reactant. A nitrogen ambient will result in a large fast state density, hydrogen in a medium to low density.

3. Post deposition annealing treatments at 700 to 1000°C in nitrogen will induce large densities of fast states in MNOS structures which are not reduced by the post-aluminum anneal treatment. A subsequent high temperature anneal in hydrogen at 700 to 1000°C of a $Si_3N_4/SiO_2/Si$ structure with a high fast state density will reduce this value back to that originally obtained in $SiCl_4$-NH_3-type deposition involving a hydrogen carrier ambient.

4. The structure of the silicon nitride layer as reflected by its etch rate will influence its ability to mask against fast-state annihilation. The faster the etch rate in concentrated HF, the less effective the masking. Deposition variables (such as reactant compositions, flow rates, and temperature) which affect nitride structure also affect the fast-state masking property.

5. The type of nitride, $SiCl_4$ or SiH_4, has a very important effect on the fast state densities. Nitrides produced by the $SiCl_4$-NH_3-H_2 reaction result in medium-low fast state densities (5-10×10^{11} cm^{-2}). Those nitrides produced by the SiH_4-NH_3-H_2) reaction, on the other hand, provide very low fast state densities ($<2 \times 10^{10}$/cm^2) but do cause significant polarization. Neither type of nitride permits fast-states induced by post-deposition anneals to be subsequently annihilated by post-aluminum treatments.

These investigators believe that for thermally oxidized silicon MOS structures, some type of hydrogen species is produced during postmetallization anneals at 500-565°C by the reaction of the field plate metal, i.e., aluminum, with water adsorbed on the oxide surface. This hydrogen then migrates to the SiO_2-Si inter-face where it annihilates the fast states although the mechanism is not known. Silicon nitride layers deposited over the thermal oxide retard this hydrogen migration, however, and any fast-states remaining after the nitride deposition or subsequent high temperature annealing treatments are not annihilated by the postmetallization anneal. A similar hydrogen species is produced during the silane-type nitrides which, at the temperatures of deposition, can migrate through the nitride layer. This causes a reduction in fast-state density, but also results in a polarization-type of instability.

Sarace and coworkers at the Bell Telephone Labs. have utilized a 400 Å silicon nitride layer sandwiched between a 600 Å thermally grown SiO_2 layer to form an alkali ion barrier for a metal-oxide-silicon field-effect transistor with self-aligned gates (a polycrystalline layer of silicon). Both n-and p-type induced-channel (enhancement) devices have been made by this process. The nitride layer serves the dual function of providing a barrier to mobile ions in the completed structure, and of acting as an etch-resistant layer during fabrication to achieve control over geometry.

Appels and coworkers at Philips (Eindhoven, Netherlands) have presented a simple technique by which some of the disadvantages of the planar technology can be overcome and new device structures made. It is based on the principle that silicon may be oxidized locally when a suitable masking pattern against oxidation can be applied. Silicon nitride is used as the masking material and is able to protect the underlying silicon against oxidation. This masking is effective as long as the nitride is not converted over its whole thickness to oxide. Because the oxidation rate of silicon is much higher than silicon nitride, thick oxides can be grown locally on the silicon, using a thin film of nitride for masking. After removal of the nitride, the remaining oxide pattern may be used for masking against impurity diffusion. Various interesting structures, e.g., well-defined thick oxide patterns countersunk into the silicon

and oxidized mesa structures, can be developed. Details on the silicon nitride
deposition and its oxidation behavior are given in the paper. The three locally
oxidized silicon (called LOCOS) structures reported by these workers are shown
in Figure 4 as an example of its versatility.

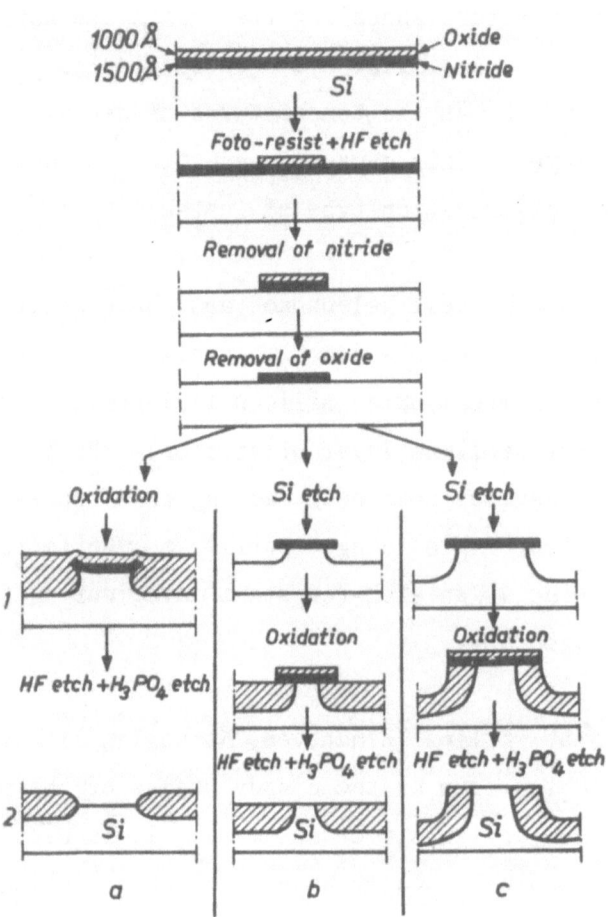

Figure 4. Process Steps in the Fabrication of
Three Different LOCOS Structures. (Appels)

(a) Oxide partly countersunk into the silicon;
(b) flat structure; (c) mesa structure.

For clearness' sake the drawings have been simplified.
In reality the structure of the window edges is more
complicated, the silicon mesas being smoothed off and
the oxide surface showing a small bump close to the
window edge.

Lawrence and Schaefer of General Electric have described recently a three-masking step MIS process utilizing a 700 Å gate oxide (grown thermally as part of the first step of the process), followed by a nitride deposition of about 500 Å. On top of this, a layer of pyrolytic oxide is deposited. Conventional processes require four masking steps.

Hillery and Clark of General Electric reported at the 1970 Electrochemical Society Meeting, on their capacitance-voltage (C-V) measurements of the masking properties of pyrolytically deposited silicon nitride films against the diffusion of alkali ion contamination. The results of their investigation indicated that the masking properties of the nitride films produced under the various deposition conditions (850-1200°C); SiH_4 concentrations from 0.003 to 0.09%) were generally acceptable. There was some indication that the film deposited using the low silane concentration (0.003%) did not mask the contamination as well as the other films produced, but even in this case the measured surface charge density was an order of magnitude lower than in the non-nitrided samples. The major conclusion of the work is that a wide latitude in process parameters can be tolerated without experiencing gross deterioration of the properties of the resulting nitride film.

The use of molybdenum films as etch masks for silicon nitride films (used as diffusion masks) has been reported by Brown and coworkers (1,2) at General Electric. The chemical stability of the refractory metal, molybdenum, allows the use of patterned films of these metals as etch masks to form high-resolution, pinhole-free patterns. Gregor (2) at IBM reported that there is no reaction between electrode metals and silicon nitride films, such as corrosion, pitting, etc., up to temperatures exceeding the melting point of the metals. Aluminum and gold adhered satisfactorily on silicon nitride. Nickel and chromium originally showed poor adherence to silicon nitride. More recent attempts have been made in depositing Cr and Ni on Si_3N_4 films. The electrodes appeared satisfactory after deposition and were sufficiently adherent to be used as etching masks for concentrated hydrofluoric acid etching of silicon nitride.

Sedgwick et al. have also studied silicon nitride dielectric films for use on germanium planar devices to solve major passivating and masking problems associated with these devices. They studied the water pickup of silicon nitride passivating films (1000 $\overset{\circ}{A}$) after humidity exposure for 46 days and showed essentially no initial water content and excellent resistance. Masking film effectiveness against gallium and indium was also studied and it was found that thick films plastically deformed the germanium. The Si_3N_4 films deposited at 800°C are in tension at room temperature and at the deposition temperature. However, a 300 $\overset{\circ}{A}$-thick film, used as a diffusion mask for gallium at 800°C, did not plastically deform the germanium wafer.

Backscattering measurements by Gyulai and coworkers at Cal Tech. have shown that outdiffusion of gallium occurs along with gallium accumulation at the surface of silicon nitride or silicon dioxide layers on GaAs substrates following anneals at 700-800°C. The investigation was made because vapor-deposited silicon nitride layers are widely used both as diffusion masks and to inhibit the effects of decomposition of GaAs during high-temperature processing steps.

Becke and White (1,2) report that depositing silicon nitride on GaAs has permitted deeper diffusions of zinc (an acceptor dopant) because of its excellent masking properties; it also does not contain oxygen, an element that produces deep donor levels in GaAs and can restrict the frequency performance of GaAs bipolar transistors. Similar improvement in transistor characteristics have been observed by Texas Instruments when the silicon dioxide diffusion mask was replaced by silicon nitride, eliminating a possible source of oxygen. For a base width of 1 micron, a gain-bandwidth product of 500 MHz was obtained.

Doo and Kerr (2) have reported on using a 600 $\overset{\circ}{A}$ silicon nitride film for partly masking GaAs substrates in zinc diffusion experiments; the silicon nitride effectively resisted diffusant penetration.

Heumann et al. made a qualitative study of the masking properties of thin (<1500 Å) silicon oxynitride films on silicon. A range of diffusion conditions were studied for doping sources including B, P, Ga, and As. Table 5 summarizes their results for the various dopants and film thicknesses. They concluded that silicon oxynitride films appear to be suitable for boron diffusion masks; however, the thickness required for masking is somewhat greater than that for pure nitride. 1500 Å will mask phosphorus for 4 hours at 1100°C; thinner films do not mask. 1500 Å will mask at 1100°C for 4 hours for arsenic diffusion. They concluded that, generally speaking, thin oxynitride films (1500 Å thick) can be used as diffusion masks for the times and temperatures suitable for shallow (<9 μ) junctions if the proper diffusant sources are used. Doo and Kerr (2) also briefly investigated the masking effect of silicon oxynitride. All the films (about 800 Å thick) which had an etch rate of 180 Å/min. or less, in ∿5% HF buffer etch, resisted steam oxidation at 1150°C for 30 minutes. When the film thickness was above 1200 Å, they resisted steam oxidation (at 1150°C for 30 minutes) even though their etch rate was 300-400 Å/min. All silicon oxynitride films resisted common silicon dopant (B, Ga, P, and As) diffusion.

Table 5. Silicon Oxynitride Diffusion Masking. (Heumann et al.)

Dopant	Diffusion system	Type of film	Thickness (Å)	Temp. (°C)	Time (hr)	Masked	Masked junct. depth. μ*	Unmasked junct. depth. μ	Comments
B	Sealed Tube	SiO_2	2000	1100	4	No	6	6	Oxide control
B	Sealed Tube	$Si_xO_yN_z$	500	1100	4	No	6	6	
B	Sealed Tube	$Si_xO_yN_z$	1000	1100	4	Yes	—	6	
P	Sealed Tube	$Si_xO_yN_z$	>1500	1100	4	Yes	—	5	
Ga	Sealed Tube	SiO_2	2000	1100	4	No	6+	6	Oxide control
Ga	Sealed Tube	$Si_xO_yN_z$	<1000	1100	4	No	3.4	6	
Ga	Sealed Tube	$Si_xO_yN_z$	1500	1100	4	Yes	—	6	
As	Sealed Tube	$Si_xO_yN_z$	1500	1100	4	Yes	—	0.6	
As	Sealed Tube	$Si_xO_yN_z$	>1000	1150	6	Yes	Spikes	1.8	Films cracked

*Where mask failure is noted but no masked junction depth is given, masking failure was determined by using a thermoelectric probe.

22

GLASS-TO-METAL SEALS

Stoller and coworkers at RCA have described a novel technique for forming glass-to-metal seals using a silicon nitride interface layer. Their method consists of depositing silicon nitride from the vapor phase onto any metal that will withstand the deposition temperature, about 700°C, in a reducing atmosphere. The adhesion of submicron-thick silicon nitride films to all metals tested appears to be excellent, and the silicon nitride was found to be readily "wet" by a wide variety of common glasses and frits. An important property of silicon nitride, one that renders it useful for this purpose, is its impermeability to oxygen and metal ions. This property is also utilized in the diffusion barrier applications against oxygen, alkali metals and semi-conductor dopants (see Diffusion Mask Section). Good seals were effected with the metals: molybdenum, copper, tantalum, Kovar, and gold and a variety of ceramic frits (Corning Pyroceram #45, Owens-Illinois #CV-102, Owens-Illinois #SG67, Corning #7570).

PASSIVATION

A great amount of interest has developed during the past five years in silicon nitride for the passivation and stabilization of semiconductor devices. This has resulted largely from the realization that silicon nitride serves as a barrier to the migration of charged ionic species, a major source of device instability. In addition, silicon nitride has a dielectric constant approximately twice that of silicon dioxide, very high dielectric strength, and can be prepared in such a way that pinhole densities are very low. Furthermore, silicon nitride passivation and metal oxide semiconductors are becoming increasingly intertwined because the nitride lowers the threshold voltage of MOS devices. Makers of bipolar integrated circuits favor the nitride principally for its protective properties.

The major function of passivating films (dielectric) is to protect the sensitive semiconductor functions from contaminants on the surface. The ability to passivate a semiconductor surface is due to actual reduction of the number of normally present surface states on a semiconductor surface. For several years now, SiO_2 has been used as a passivating layer over semiconductor devices. For silicon, these films have generally been prepared by the thermal oxidation of the silicon wafer in oxygen or water ambients at temperatures ranging from 900 to 1200°C. Other methods of producing passivating layers of silicon dioxide, both on silicon and other semiconductor materials, include pyrolytic decomposition of oxysilanes, vapor phase reactions of silicon and oxygen-containing compounds, sputtering, plasma oxidation and anodization.

The use of silicon nitride for passivation and encapsulation purposes will become more attractive when the deep-seated nitride traps can be eliminated, according to Kendall, as a result of his C-V and I-V measurements and studies of trapping mechanisms. He advocates the use of ultra-clean fabrication to produce a trap-free nitride since the presence of impurities causes local crystallization which in turn appears to contain the deep traps. The presence of oxygen is also a problem since even small quantities lead to the formation of silicon oxynitride which is orthorhombic in structure and is often mistaken for a third (γ) phase of silicon nitride.

Myers at the IIT Research Institute Reliability Analysis Center has made a state-of-the-art survey of silicon nitride surface passivation usage by discussing:

a) Problems with the use of silicon nitride
b) The uses of silicon nitride in the planar process
c) Silicon nitride processing
d) Reliability experience with silicon nitride

M.P. Lepselter at the Bell Telephone Laboratories (in 1966) was the first to investigate the use of silicon nitride as a sealing technique to encapsulate thousands of silicon semiconductor devices while they are still on a single silicon wafer. The junction seal is formed by application of a layer of silicon nitride and beam lead contacts to the silicon dioxide layers of the transistor and integrated circuit device. Deposition of the silicon nitride layer is accomplished by mixing two gases, silicon hydride (silane or SiH_4), diluted in hydrogen, and ammonia near a heated silicon slice. The silicon nitride adheres to the slice to form a protective barrier against penetration of sodium and other metallic ions. Beam lead contacts are then applied to the device as shown in Figure 5. The beam lead contacts form a strong mechanical bond with the silicon nitride layer, thus sealing the required contact area against ion penetration and preventing the leads from becoming detached. The completed wafer is then finally etched apart. The individual devices and circuits are impervious to acids that would ruin canned devices, and can withstand accelerations of 300,000 G.

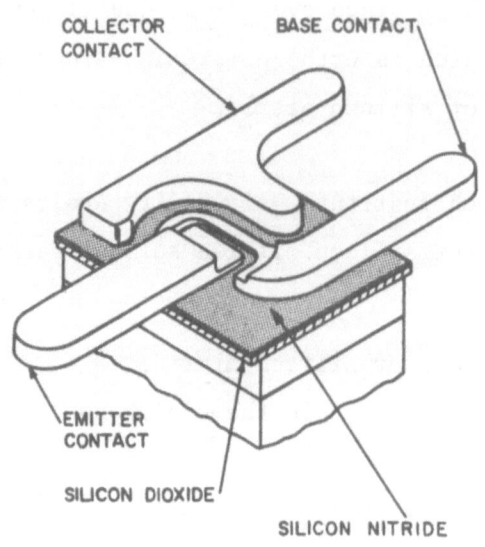

COLLECTOR CONTACT

BASE CONTACT

EMITTER CONTACT

SILICON DIOXIDE

SILICON NITRIDE

Figure 5. New Junction Seal Shown on
Epitaxial Silicon Transistor.
(Lepselter)

High-power mesa-type rectifier devices have been experimentally passivated
with SiO_2/Si_3N_4 layers: (a) A SiO_2 layer is prepared by deposition onto the
junction surfaces at 600°C by the reaction of silane in a N_2 carrier gas
(thickness of 500 Å at a deposition rate of 100 Å/min). (b) A Si_3N_4 layer is
deposited on this by the reaction of silane and ammonia in N_2 at a temperature of 700°C (thickness of 2500 Å at a rate of 120 Å/min). Reverse I-V
characteristics, high-temperature yield, and blocking life tests were run
on these SiO_2/Si_3N_4 passivated devices and compared with polyimide coated
specimens. It was found that the high-temperature (175 and 200°C) performance
of the oxide-nitride passivated units was superior to the polyimide coated
units. Figure 6 shows how the two coated processes compared with respect to
leakage current at 175°C. These studies carried out by Verderber and coworkers
at Westinghouse Electric Corp., Semiconductor Division, indicated that a
strong correlation between the high-temperature leakage current and thermal
stability of the device existed. The lower the leakage current for a device
at 175°C, the more reliable it will be during operation at high temperature.

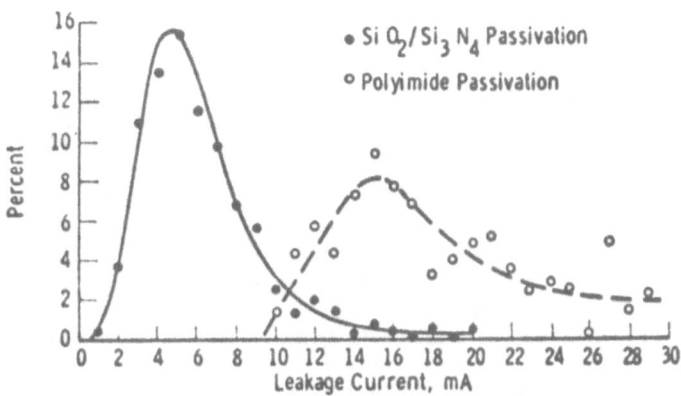

Figure 6. Frequency Distribution of Units Having a 175°C
Junction Temperature Leakage Current Within a
1-mA Range at 2000 Volts. (Verderber et al.)

Gruber and Verderber of Westinghouse at the 1970 Electrochemical Society
Meeting in Los Angeles reported on their investigations of the effects of
film thickness and deposition temperature on SiO_2/Si_3N_4 passivated high
voltage mesa diodes. The units were passivated with successive layers of
the oxide and nitride using chemical vapor deposition (CVD) techniques with
O_2, NH_3, and SiH_4 as reactants. The deposition temperatures were studied
over a range of 600 to 700°C for the SiO_2 deposition and from 650 to 750°C
for the Si_3N_4 deposition. Thicknesses of both SiO_2 and Si_3N_4 were varied
from 0 to 3000 Å. Their results showed that effective passivation can be
obtained with the SiO_2/Si_3N_4 system for a wide range of both oxide and
nitride thicknesses and deposition temperatures. The SiO_2 thickness should
be between 100 and 1000 Å while the Si_3N_4 layers are best within the 1500 to
3000 Å thickness range (thicknesses greater than 3000 Å could result in
cracking of the Si_3N_4 films). Deposition temperatures must be below 650°C
for the SiO_2 deposition and below 750°C for the Si_3N_4 deposition in order
to minimize leakage currents.

International Rectifier Company has announced recently that it has adapted a
semiconductor planar process to making power devices using silicon nitride

27

and glass passivation. This allows the use of unpackaged SCR (silicon controlled rectifier) functions. The device or rather the SCR junction is
mounted on a tungsten or molybdenum buffer plate, and the latter mounted to
the metal lead frame. The active devices in this hybrid assembly are encapsulated in a high-temperature, moisture-resistant epoxy.

In 1968, Dalton and Drobeck studied the movement of sodium in silicon nitride
films by means of radioactive sodium. The sodium contaminated samples were
treated in two ways: (a) heated without a field, and (b) with a field
applied to enhance the sodium motion. Autoradiographs were made to follow
the diffusion and drift experiments at 400°C. These investigators found the
ability of the sodium to penetrate the silicon nitride films was related to
the crystalline size of the film. Sodium was found to diffuse more easily
into the films composed of larger crystallites and yielded the data shown
in Figure 7.

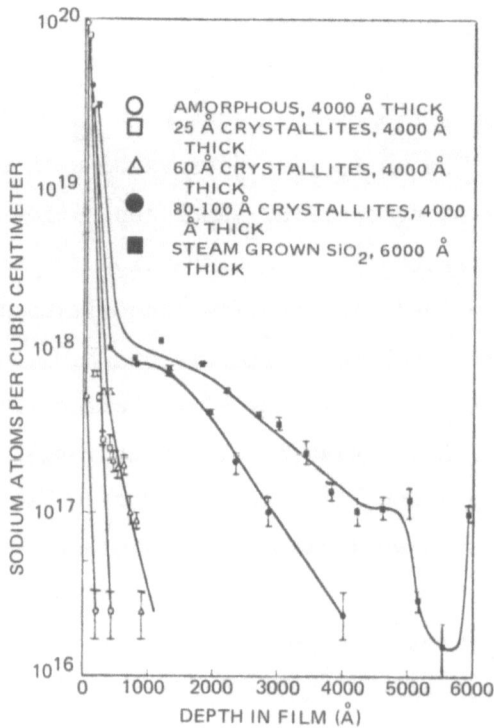

Figure 7. Sodium Diffusion Profiles in
Silicon Nitride and Silicon
Dioxide. (Dalton & Drobeck)

28

No field enhancement of the diffusion of sodium was detected in any of the silicon nitride films studied, regardless of crystallite size or type. Schmidt and Wonsidler noted from the preceding Dalton and Drobeck paper that poly-crystalline silicon nitride films are undesirable because they are not effective barriers to sodium migration.

Trapp and Preece investigated various silicon nitride passivated integrated circuits and showed that they exhibited exceptional stability to inversion after reverse bias accelerated stressing up to 1000 hours at 175°C. They presented stress data as a function of the process used and showed that the limiting factor to device stability is the cleanliness of the SiO_2 layer.

Swann et al. fabricated MIS transistors from a 1000 $\overset{\circ}{A}$ layer of silicon nitride (deposited using 3:1 NH_3/SiH_4, onto 2 ohm-cm-n-type Si). The threshold voltage for the subsequent p-channel devices was -4.5 V, G_m (at 1 ma I_D) = 500 μmhos and hole mobility = 110 cm^2/V-sec. The I-V characteristics of the source and drain junctions indicated diodes comparable in performance to those passivated with thermal SiO_2. The results, coupled with the absence of sodium migration, indicate that the low temperature nitride is suitable for passivating silicon planar devices.

Schneer et al. have conducted tests on sealed-junction transistors consisting of $PtSi_2$-Ti-Pt-Au contacts and a silicon nitride overcoat, to determine and compare its reliability with standard silicon planar transistors sealed in a vacuum-tight enclosure. Figures 8 and 9 show the enhanced reliability of the nitride types.

Hartman and Herr (1) have noted that the impermeability of silicon nitride to alkali ions, water vapor, and even hydrogen makes it most attractive as an outer barrier before contact formation on silicon planar structures. On the other hand, the electrical properties of silicon nitride films have been highly variable. In general, silicon nitride films are inferior to SiO_2 as an electric insulator. Therefore, they have not been incorporated as a passivation or masking layer at the silicon interface. Silicon nitride thus has an application potential as a barrier between the SiO_2 and the metal forming system of the semiconductor structure.

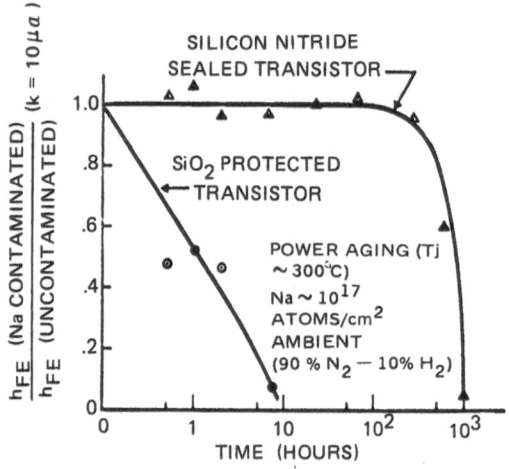

Figure 8.   The Effect of Sodium Contamination on The Gain of Sealed-Junction and SiO_2-Protected Transistors. (Schneer et al.)

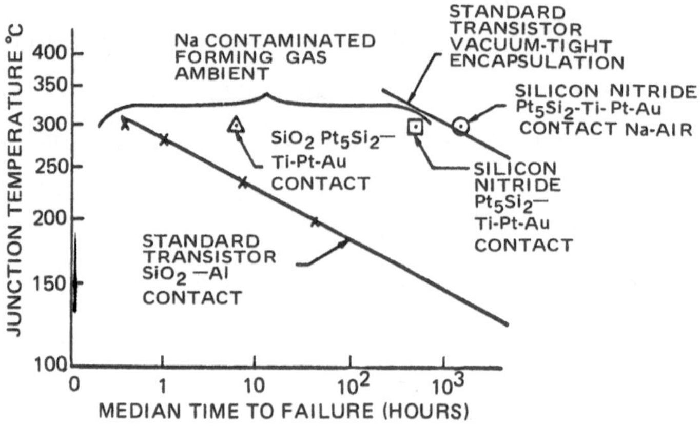

Figure 9.   Sealed-Junction Reliability. (Schneer et al.)

Bell Telephone Laboratories is credited with discovering that silicon nitride passivation would work on MOSFETs while Motorola claims it was the first to introduce it as a standard production technique for all discrete lines. A 1968 review article in ELECTRONICS quoted Motorola engineers as saying silicon nitride gives a far more uniform passivation layer than SiO_2 permitting them to specify a gate-to-source breakdown of 50 volts for a typical device, compared with 15 volts for a MOSFET passivated with SiO_2. Further, some of their new components had completed 1000-hour life tests. According to Barney, a number of companies have explored silicon nitride passivated devices and found they tend to leak current. Moreover, silicon nitride does not interface well with silicon; the uncommitted bonds result in surface states that look like uncontrollable charges. Threshold voltages on devices made with a

30

simple nitride insulator, shift dramatically in one direction when the devices are biased negatively, and in the other direction when they are biased positively. To prevent this shift, it is necessary to put a layer of oxide on top of the silicon, and a layer of nitride on top of the oxide.

Recently, Raytheon has announced the release of silicon nitride passivated beam lead bipolar integrated circuits similar to those actively developed by Bell Telephone Laboratories. In such devices, the nitride film essentially forms a conformal hermetic package. Legat has detailed this development as shown in Figure 10 which indicates the utilization of silicon nitride as a passivation layer in conjunction with PtSi, Ti, Pt and Au.

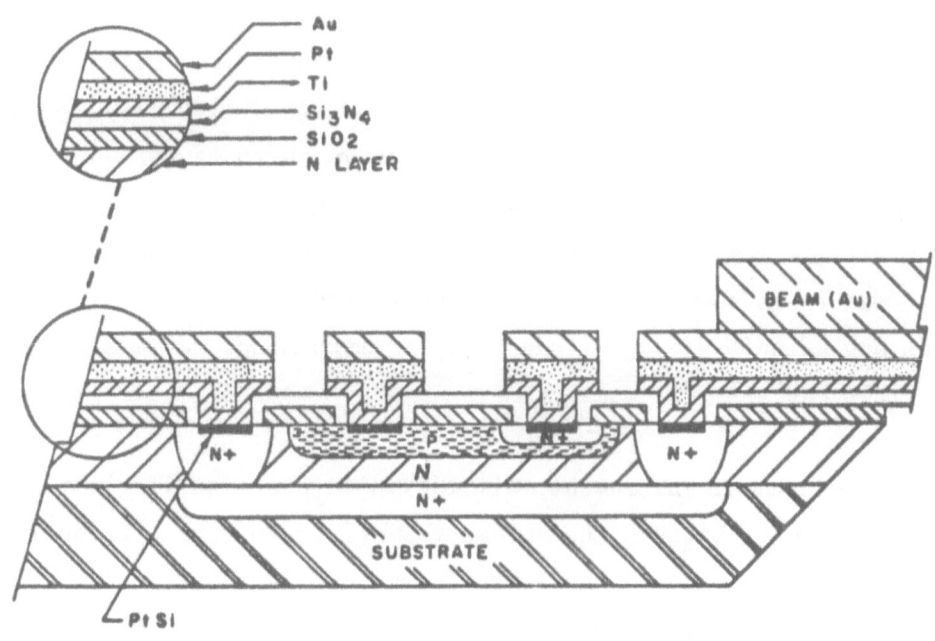

Figure 10. Cross Section of Beam Lead
Device with Interconnections.
(Legat)

31

The use of silicon nitride to make beam-lead, sealed-junction, monolithic, integrated circuits has been described by House and Whitner of Western Electric Company. While the entire processing is quite complex involving wafer preparation, buried-layer diffusion, epitaxial-layer deposition, isolation diffusion, deep-collector diffusion, base diffusion, emitter-diffusion, contact window-deposition, contact-window opening, platinum silicide contact formation, titanium deposition, platinum deposition, gold intraconnection deposition, gold beam-lead deposition, and circuit separation, the silicon nitride layer improved the reliability and reduced the cost, increasing the overall yield. The silicon nitride is deposited to a thickness of about 0.3 micron onto the wafers, which are maintained at a temperature of 880°C. It is made by the pyrolytic reaction of silicon tetrachloride and ammonia. The etching of windows through the nitride is accomplished with boiling phosphoric acid using an oxide layer on top of the nitride.

In October 1967, the General Instrument Corp. announced that it was producing the industry's first silicon nitride passivated computer diodes. Such high-speed silicon nitride diodes, produced for computer applications eliminate common modes of diode failure. Passivating with silicon nitride rather than with silicon dioxide provides the units with high reliability and eliminates unstable reverse breakdowns, excessive leakage currents and contamination during chip handling and packaging. The company states that the use of the silicon nitride layer, impervious to the movement of sodium ions and chemically inert, results in good reliability. The nitride process permits batch sealing and the use of soft glass, with no loss of mechanical reliability or increase in assembly cost.

According to General Instrument, nitride units are less liable to adverse channeling effects because the volume resistivity of the nitride is lower than that of oxide, so that negative charge can leak off rather than accumulate on the surface; the required nitride layer is only 1/8 as thick as the usual oxide, enabling the charge to leak easily. The device which is packed

in miniature DO-35 or DO -7 packages for computer applications, is shown
schematically in Figure 11.

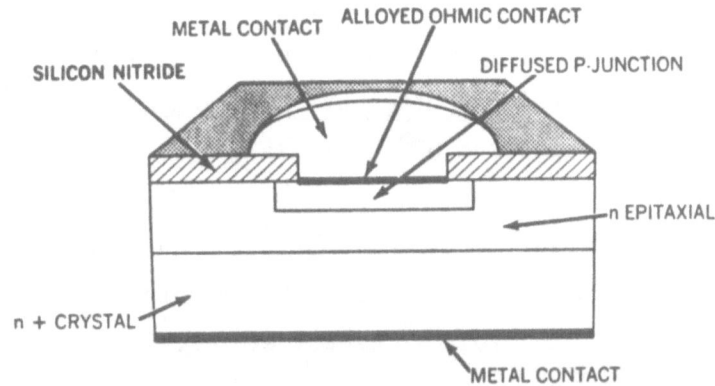

Figure 11. Cut-Away View of Silicon Nitride
Passivated Computer Diode.
(General Instrument Corp.)

In November 1970, they announced the availability of GIANT (General Instrument Advanced Nitride Technology) MOS shift registers in plastic DIPs (Dual-
in-line Packages). A wide range of dynamic and static shift registers are
available: 512-bit, 256-bit, 200-bit, 100-bit, and 16-bit; also in dual and
quad forms. These devices are completely compatible with TTL/DTL and MOS,
both electrically, through the use of low threshold nitride processing, and
mechanically, through the use of the 14-lead plastic DIP. The unique combination of the passivating nitride with low-cost plastic encapsulation brings
the price down to attractive levels.

Gray and coworkers at Sprague Electric Co. have presented reliability data
on silicon nitride shields for plastic encapsulated transistors at a recent
Reliability Symposium. It is well known that plastic epoxy encapsulated
transistors are vulnerable to ionic contaminates such as sodium, especially
in the presence of moisture which penetrates any plastic material and causes

sodium to migrate rapidly. They found that silicon nitride films deposited over passivation SiO_2 films shielded these passivation layers from both ionic contaminants and moisture. Radio tracer measurements of sodium-22 and initiated water demonstrated that even under conditions of some external contamination, high humidity and high temperature, silicon nitride reduced the rate of sodium transport into the passivating layer by a thousand fold. Doo and Kerr also reported on their evaluation of silicon nitride and oxide passivated diodes conducted at IBM for NASA in 1968. Little effect was found with intentional sodium chloride contamination as compared with non-contaminated devices after being stressed 666 hours at 175°C.

In 1968 Westinghouse Molecular Electronics Division announced their new process that hermetically seals integrated circuits at the chip level and is known as the "Goldilox" approach. It involves three features:

a) Silicon nitride passivation

b) Glass over interconnections

c) Titanium-gold bonding system

The silicon nitride passivation replaced the standard lead oxide or phosphorus glass. It significantly retards ion migration, eliminating inversion and I_{DR} and I_{CEX} leakages at high temperatures. According to an advertisement in an electronics journal, Goldilox chips are hermetically sealed within themselves and have been successfully operated under water without packaging for periods of over one hour. Figure 12 depicts the structure involved in the process.

Silicon nitride films have also been utilized in the dielectric air isolation scheme identified as "Decal" and developed at RCA for fabricating integrated circuits. Here, the tungsten metallized device wafer is coated with a silicon nitride layer to provide an impurity barrier. The wafer is then bonded facedown to a glass substrate by pressure fusion. The excess silicon is removed in a series of operations from the back side of the device wafer, according to Duffy and Kern, leaving only islands of silicon and exposed tungsten bonding pads on the glass, all of which are bonded to the glass via silicon nitride.

ITT's Semiconductor Division has introduced a nitride-passivated circuit, a dual 16-bit static shift register according to Barney.

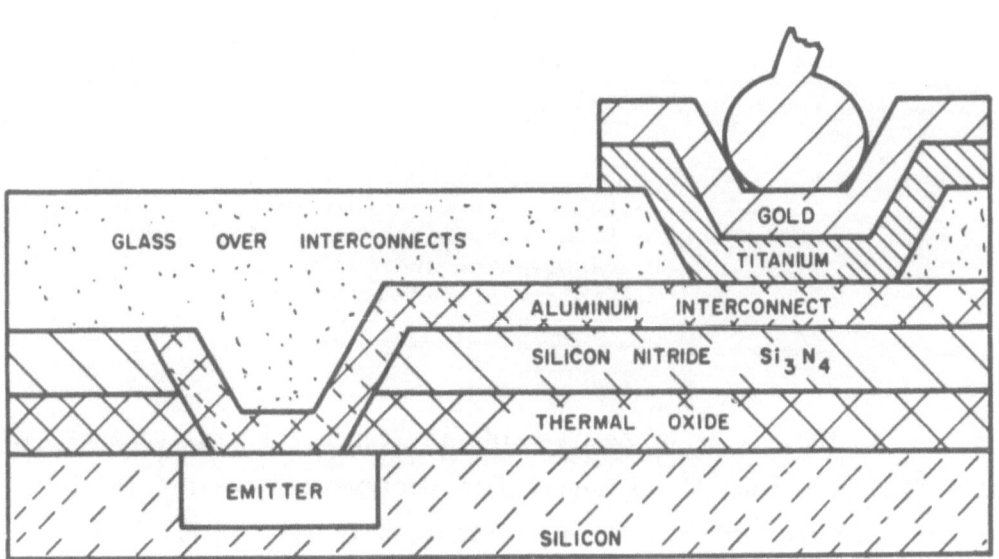

Figure 12. "Goldilox" Hermetically Sealed
Structures. (Westinghouse)

Fairchild Semiconductor Corp. in 1969 reported an eight-fold reduction in failure rate per 1000 hours in a reliability study conducted on nitride-passivated diodes. Figure 13 illustrates the silicon nitride passivation configuration used by Fairchild in their FDH-6 chip:

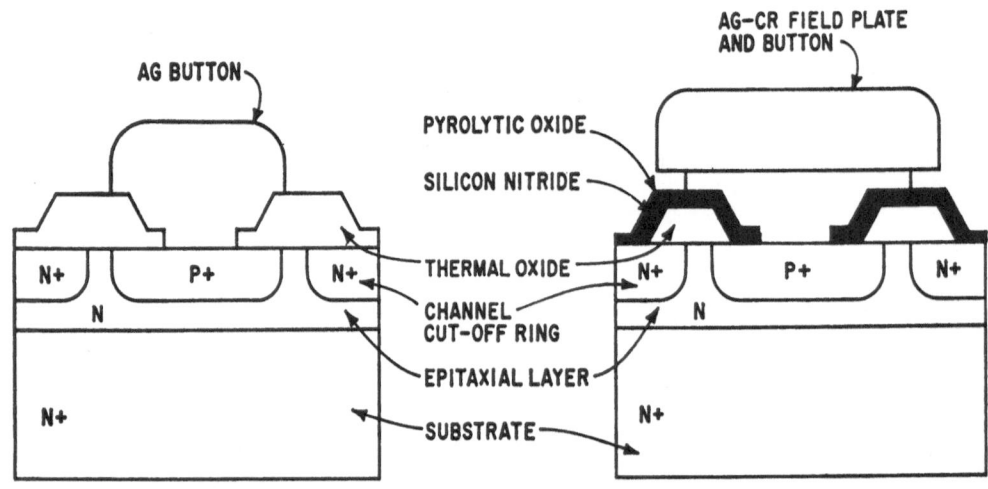

Figure 13. Fairchild FDH-6 Chip Before and After the Switch to Nitride Passivation and Addition of a Field Plate. (Fairchild)

The inherent instabilities of metal-oxide semiconductor field-effect transistors (MOSFETS) have made it unpopular with designers, however, the Semiconductor Products Division of Motorola Inc. reported in 1968 it had overcome MOSFET instabilities by using silicon nitride passivation. The result has been greater yields and volume, permitting marked device cost-reduction. By 1968 Motorola had converted to a mass production line status, its silicon nitride-treated devices and found them more stable than silica-passivated transistors.

Electrically, their nitride-passivated units are identical replacements for the conventional devices and are completely interchangeable with them. However, the company said the new devices carry with them a higher allowable gate voltage that results from the greater uniformity of the silicon nitride passivating layer. Their process uses the standard epitaxial deposition apparatus. Silicon wafers are heated in a quartz tube at about 1000°C. Hydrogen, the carrier gas, flows through and contains the silane and ammonia, which react to deposit the silicon nitride. They note that conditions are very critical. The main stream flow rate is 40 liters per minute, while the silane flow is 3 cc per minute and the ammonia flow is 100 cc per minute. If the silane flow goes up to about 6 cc per minute, a film forms that cannot be etched at all. The thickness of the film can be controlled within 5% accuracy with a deviation from uniformity of 2%. With other investigators, they found that they had to lay down a thin silica film on the silicon substrate before the silicon nitride, otherwise, a transition zone forms between the nitride and silicon that gives bad electrical properties and is very difficult to etch. The improvement of these new semiconductors over the old ones was striking to Motorola. It was common for the old MOSFETs to have a threshold voltage that varied by 6 V in less than 100 hours of testing at 150°C. By contrast, it was extremely rare to find nitride MOSFETs whose threshold voltage varied by more than 600 mV after 1000 hours of the same test at 200°C. Some of their advertized products include the following:

1. MFE-3002; n-Channel silicon nitride passivated MOS field-effect transistor.
2. MFE-3003; p-Channel silicon nitride passivated MOS field-effect transistor.
3. 3N155, 3N155A, 3N156, 3N156A; p-Channel silicon nitride passivated MOS field-effect transistors.
4. 3N157, 3N157A, 3N158, 3N158A; p-Channel silicon nitride passivated MOS field-effect transistors.

Applied Materials Technology, Inc. developed (in 1969) two horizontal chemical vapor deposition reactors that uniformly deposit either silicon nitride or silicon dioxide--or both--on semiconductor wafers. The firm points out that the reactors are not intended for customers putting down oxide passivation layers alone, but for those depositing oxide-nitride sandwiches or nitrides alone and uses reactors with infrared heat sources, consisting of a series of high-intensity mercury halogen lamps. The total passivation reactor system will deposit oxides from silane and carbon dioxide at 900°C, or nitrides from silane and ammonia at 700° to 850°C to form the usual oxide-nitride sandwich employed with MOS devices described above.

The use of silicon nitride for passivation against oxygen and hydrogen diffusion during processing steps in germanium planar technology has been reported by Sedgwick and coworkers at IBM. The silicon nitride was prepared by reacting NH_3 and $SiBr_4$ in forming gas (at 800°C, 300 $\overset{\circ}{A}$/min) or in N_2 (at 700°C, 150 $\overset{\circ}{A}$/min).

ISOLATION

In 1967, R.P. Donovan at Research Triangle Institute made a survey of
isolation methods used in integrated silicon device technology (under an
Air Force Contract), and found the following methods available: resistive
isolation, junction isolation, dielectric isolation, isolation by silicon
oxide and other dielectrics such as silicon nitride. Each approach has its
merits and disadvantages in cost economics and trade-offs. The advantage of
the dielectric method over the reverse biased junction is that it exhibits
no polarity dependence and separates the two areas electrically. The dielectric reduces leakage currents between adjacent components by a factor of
about 10^4 over that obtainable by junction isolation; it further reduces the
capacitive coupling by an order of magnitude. Their survey included the
various methods of preparing silicon nitride as were reported in Volume 3
of this Handbook Series. the instabilities of the silicon nitride (in MNS
structures) are mentioned, which, moreover, could be controlled by its
combination with silicon oxide as a laminate structure. Mention is made of
Lepselter's use of silicon nitride in beam-lead devices.

Jones and Doo have depicted as shown in Figure 14 the fabrication procedures
involved in making complete isolation structures using silicon nitride:

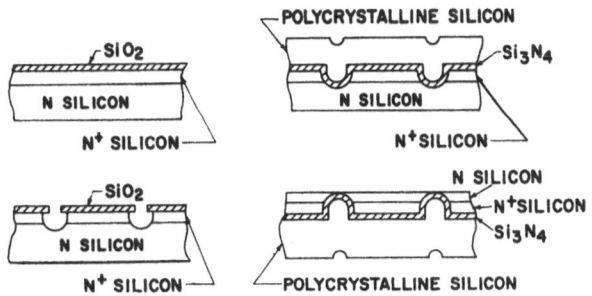

Figure 14.   Fabrication Steps for Complete Isolation
             Structures Using Silicon Nitride.
             (Jones and Doo)

Starting with n-type Si wafers, an N^+ epitaxial layer is deposited. This layer will eventually be the "subcollector" for a transistor structure. The wafer is oxidized and opened using photolithography steps, then etched to form channels or moats. After the moats are formed, the oxide is stripped and the desired thickness of silicon nitride is deposited. A thick layer of polycrystalline silicon is then deposited to form the support for the isolated islands. The final step consists of etching back the original wafer until the silicon nitride (Si_3N_4) humps emerge through the surface and completely isolate the remaining silicon.

Doo (1) has claimed that Si_3N_4 is useful in providing a dielectric isolation medium for semiconductor devices.

MEMORY DEVICES

The greatest advance has been made recently in the theoretical understanding of the memory device, according to Soref of Sperry Rand Co. This is based on the existence of two coplanar regions having mutually independent conductance laws. Early in the exploration of silicon nitride films in MIS systems, it was found that they exhibited a charge storage property and a substantial effort has been devoted to measuring and explaining this phenomenon. A semi-quantitative picture is the best available at present. Since the stored quantity consists of charges in the dielectric of the gate of the insulated-gate field-effect transistor (IGFET), the switching characteristic deals with the amount of charge stored. The study of write-in time is concerned with a determination of the rate of charge transfer. The persistance of information really depends on how long the charges are stored. Various models have been put forth to explain the charge transport mechanisms, i.e. charging and discharging times in the various MIS structures (including double layer MI_1I_2S structures). The theoretical work ties all of these threads together in a model of the movement of charges in a solid state device as a function of applied voltage and the physical and geometrical parameters in a qualitative way and awaits detailed measurements for complete confirmation.

As indicated in the MNS Capacitor Section, Sperry Rand Research Center investigators (see Wegener (1)), explored the effects of pulse duration and pulse amplitude (voltage) of MNS capacitors which indicated or prompted the fabrication of MNS variable threshold voltage transistors (MNS-VTT) having a great potential for memory elements that store information in a nonvolatile manner. Such a silicon nitride IGFET can be regarded as an interacting combination of an MNS capacitor and two reverse biased p-n junctions. In particular, the same phenomena which determine the inversion voltage of the capacitor also determine the turn-on or threshold voltage of the transistor. The hysteresis and storage effects observed in MNS capacitors were also found to occur in the MNS transistors. Subsequently, the threshold characteristics of these transistors were investigated as a function of various pulse conditions with the results shown in Figures 15 and 16.

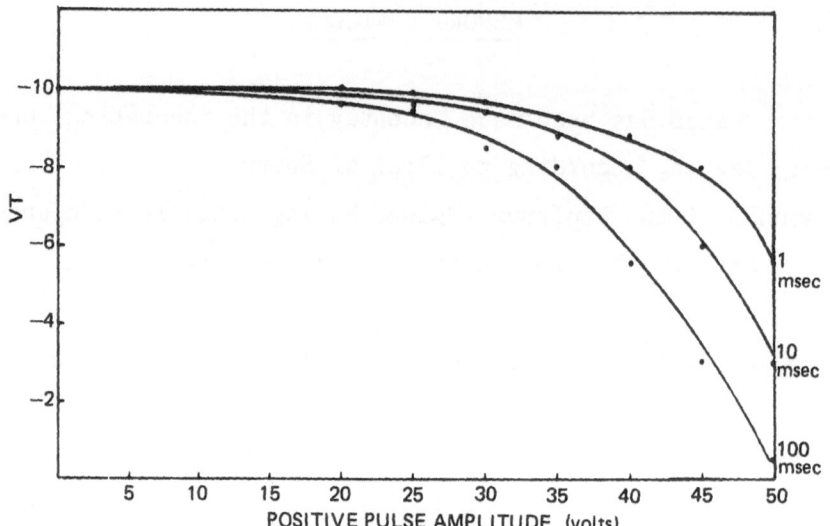

Figure 15.   Threshold Voltage V_T of MNS-VTT as a
Function of Positive Pulse Amplitude
and Length.   (Wegener (1))

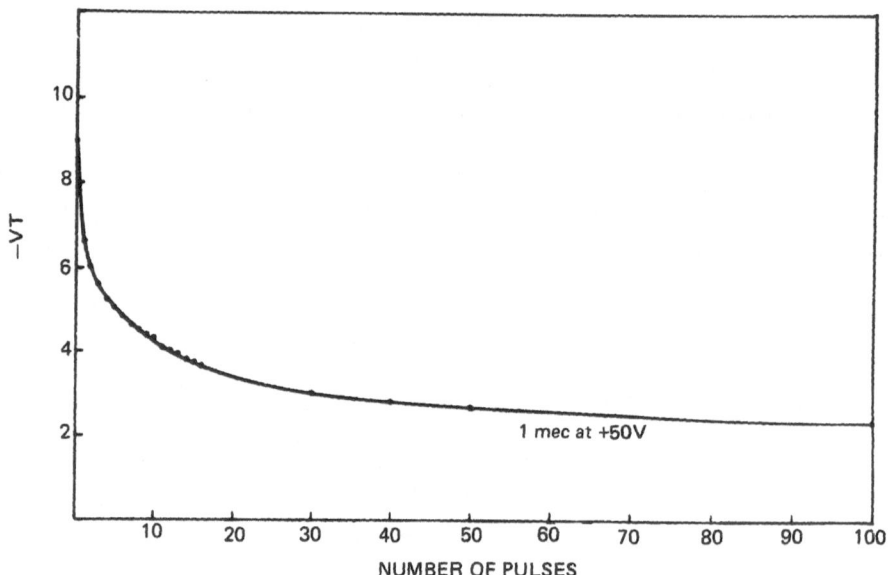

Figure 16.   Threshold Voltage V_T of MNS-VTT as a Function
of the Number of Positive Pulses of +50 V
Amplitude and 1 msec Length.   (Wegener (1))

Life tests as shown in Figure 17 were also performed in order to check how well these devices would stand up under numerous switchings from a high to a low threshold voltage (V_T).

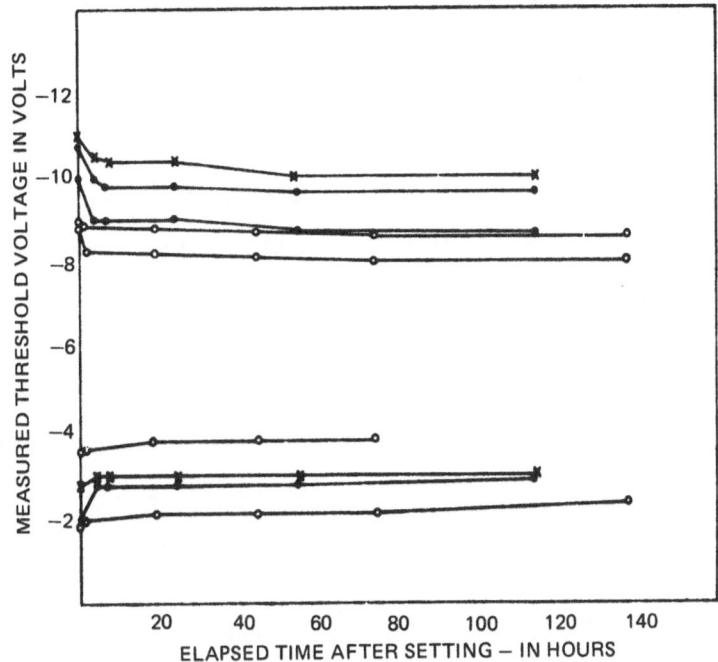

Figure 17.   Persistence of Stored Information: o 25°C,
no bias on gate; x +125°C, no bias on gate;
● +125°C, constant -5V bias on gate.
(Wegener (1))

From these data, it could be extrapolated that digital information stored in such devices can persist for many days under operating bias conditions and at temperatures up to at least 125°C. Life tests have indicated unchanged operation for several million switching cycles. Thus, Sperry Rand Research Center workers conclude that if one were to use the MNS-VTT in a stored program computer, reprogramming one hundred times a day would permit the use of even the present devices for a minimum time of the order of 200 years.

In discussing the memory mechanism of the MNS-VTT device, Wegener (1) notes it is the induced change in threshold voltage which contributes the memory effect. This change is due to the storage of charge in the gate dielectric

43

of the memory IGFET. Thus, gate structures and gate materials become important parameters. The exact location of these charges was of importance for the understanding of the device and for the control of its fabrication. Laboratory experiments established that the charges were located within 100 $\overset{\circ}{A}$ of the insulator-silicon interface. In addition, experiments demonstrated that all the charge observed was transferred by the conduction current, and vice versa--hence, a study of current characteristics assumed greater importance. It was also found that the charge transfer mechanism is independent of temperature, at least between 25°C and 100°C. This suggests a mechanism involving either tunneling of conduction electrons through a barrier or tunneling of electrons from isolated states into a conduction band. The temperature independence of the current pulse is further evidence that the positive charges in the dielectric were formed by electrons leaving neutral states. The density of neutral states in the dielectric is clearly independent of temperature.

In order to understand the charge mechanisms, the J-V curves and conduction laws must be properly interpreted. The current density, J, obtained for an applied field, E, is represented by the following three contributions:

$$J = J_1 + J_2 + J_3$$

where J_1 is the current arising from an internal Schottky emission of trapped electrons into the conduction and is known as the Poole=Frenkel effect. J_2 is the current arising from the internal field emission of electrons from traps into the conduction band. It is essentially a tunneling effect. J_3 is the current arising from the hopping of thermally excited electrons between isolated states. It has an ohmic characteristic, but is dependent on a thermal activation energy.

With this previous background discussion in mind, Wegener [1] states that there are two basic features of the memory effect. The first is the requirement of a dielectric with at least two regions of different conductivity. This structure results in the storage of charge at the interface regions in any combination of dielectrics. The second requirement is that at least one of the two layers should exhibit a highly nonlinear conductivity, very high at high fields and very low at low fields. This property results in short write-in times during the application of high fields and long storage times under the normal conditions of low field. An analysis of the conduction laws in each layer with respect to the bias voltages shows that both positive and negative applied voltages can result in both positive and negative charge accumulation at the insulator-insulator interface, depending on the individual conductance laws in each region (I_1 and I_2). Figure 18 summarizes these bias conditions.

Finally, whatever the sign of the charge accumulating at the interface, it will give rise to a field which opposes that in existence in the more highly conducting region. The field due to the accumulating charge will at the same time add to that in existence in the less highly conducting region. All charge accumulation will stop when the effective field across the high conductance region has been lowered to such an extent that its current density is equal to that passing through the low conductance region, whose field has been increased by accumulated charge. The end point of charge accumulation is therefore obtained when the conduction currents across both regions are equal.

A "learning machine" was constructed with MNS-VTT memory transistor (Fig. 18) as part of NASA contract (NAS 12-570, Wegener (1) and Lewis (1)) by Sperry Rand Research Center. This adaptive circuit utilizes the controllable threshold voltage of the MNS-VTT to determine its electrical output. The setting of the threshold voltage is influenced by "reward" and "punish" stimuli. The theory of this circuit application has been partially established. It was shown that learning can be accomplished even by an imperfect trainer. The only requirement of the trainer is that he make more correct decisions than incorrect decisions.

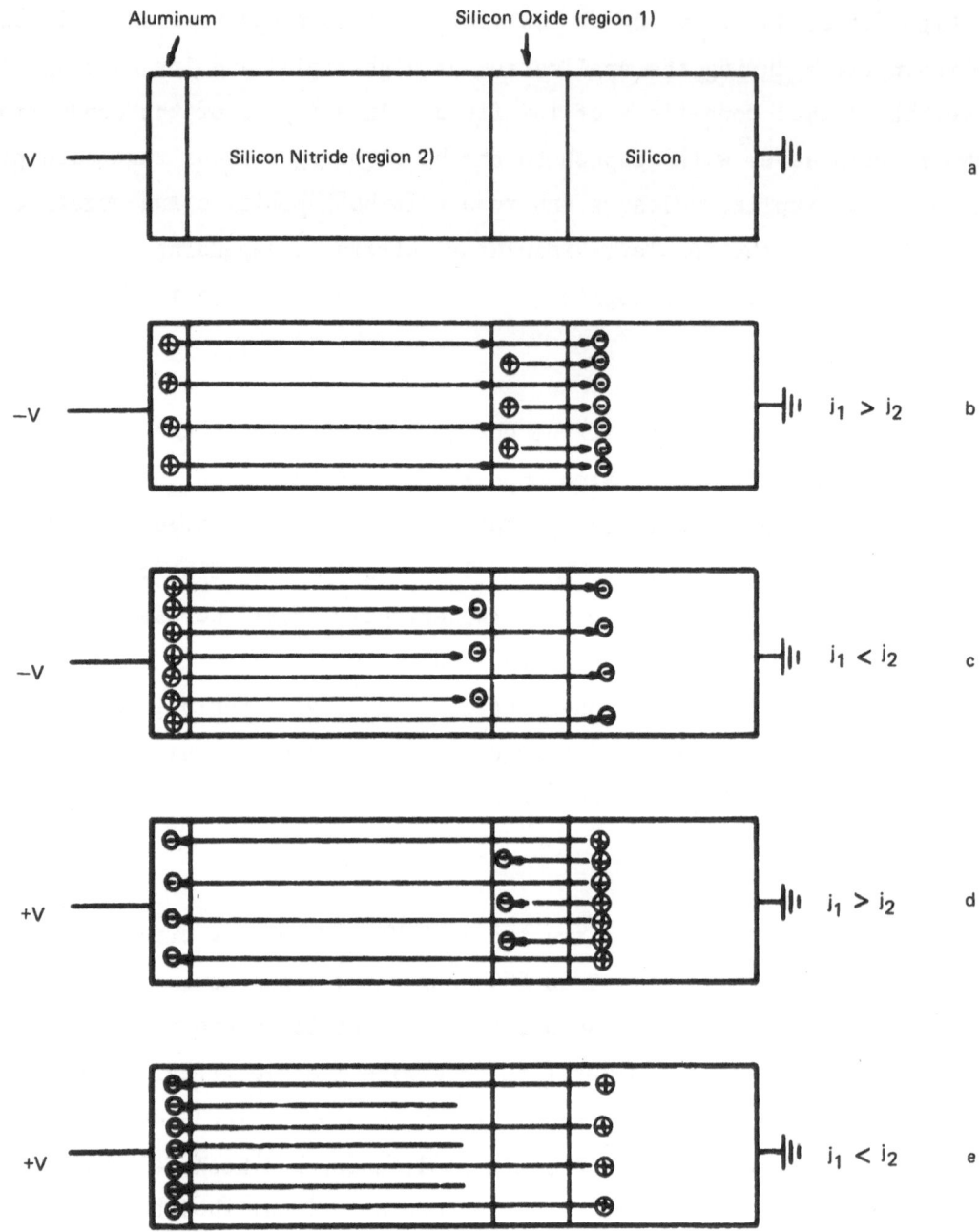

Figure 18. Effect of the Polarity of the Applied Voltage
and the Relative Magnitude of j_1 with Respect
to j_2 on the Polarity of the Charge Sheet
Built Up at the Insulator-Insulator Interface.
(Wegener (1))

46

Weiss discussed a MNS memory device which is structurally similar to the p-channel MOS device family. A silicon nitride material is used that permits a semi-permanent change in the threshold of the device. The threshold is changed by applying a large voltage to the gate and then operating below this voltage level. The time constant at which the threshold of the device drifts to its original level is of the order of years and is dependent on how long the large voltage is applied to the gate of the device. The significant benefit gained by going to this structure is non-volatile semiconductor storage. The device as shown in Figure 19 is still in the laboratory stages, but offers a means of producing very high density non-volatile random access memory and electrically alterable read-only memory.

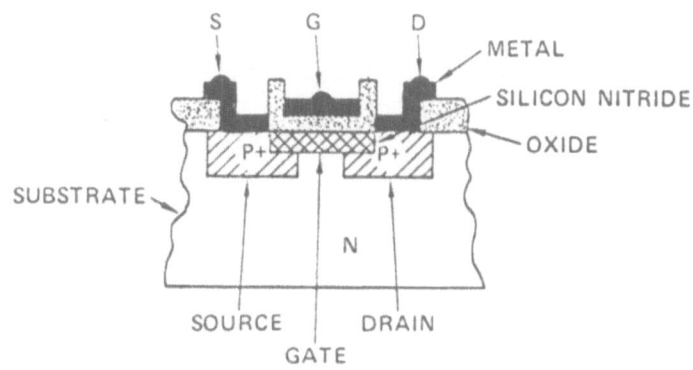

Figure 19. MNS Transistor.
(Weiss)

To write information into this device, a large voltage is applied to the gate. This change has a long time constant of days to years depending on the duration of the write pulse. To read the state of the device, a voltage lower than the write voltage but higher than the original threshold is applied to the gate. If the device is in the high threshold state, it does not turn on; if it is in the low threshold state, it does. The device (laboratory stage) is fabricated on 6 square mils of chip real estate. Access time is about 100 nanoseconds in an array. Write time is several microseconds.

An MNS light sensitive memory element was described recently by Sewell (1) of Sperry Rand Research Center, that performs the functions of both light detection and memory. It is based on the previously described variable threshold MNS-FET developed by Sperry Rand. In this device (under the proper conditions), the rate at which the FET gate is charged with a writing pulse of given amplitude, depends upon the intensity of the illumination incident on the device. In an array of such devices, therefore, selective illumination of certain regions during the writing cycle will implement charge storage, while in the nonilluminated area writing will be inhibited. This behavior is immediately applicable to image and character storage and can provide a direct optical input to a memory array. In this device, charge storage is accomplished in a two insulator memory transistor by transporting carriers to the insulator-insulator interface under the influence of an applied voltage, the magnitude of which determines the rate of charging, or the writing speed of the device. However, by allowing the source and drain of the FET to float during the writing pulse, a transient space charge layer is formed in the semiconductor, thus reducing the applied voltage that appears across the insulator. The field across the insulator is then reduced so much that only negligible carrier transport may occur. Physically, the charging pulse (of proper polarity) initially depletes a volume of the semi-conductor of majority and minority carriers, leaving the ionized donor or acceptors to form the space charge region. Minority carriers, however, are generated in the space charge region resulting in the formation of an in-creasingly large inversion layer and a diminishing space charge width. Thus, the duration of time that the voltage on the insulator is reduced by the space charge voltage depends on the generation rate of minority carriers. The effect of the illumination is to increase this generation rate and thus decrease the delay time for writing. Experimentally, Sewell observed as much as four orders of magnitude delay between the illuminated and non-illuminated state and these delay times agree accurately with the calculated values.

The semiconductor memory technology is presently arriving at a strong competitive position with magnetic core memories, as they offer substantial advantages in most performance categories. However, they cannot provide a substitute for the non-volatile storage capability (retention of stored information without an external power source) of magnetic memories. Most proposed non-volatile semiconductor storage devices rely on charge storage in a dielectric which forms part of the gate of a field effect transistor (FET). Most of the attention has focused on silicon nitride, according to Frohman-Bentchkowsky (1), with particular emphasis on the MNOS structure. The storage function in a MNOS transistor is achieved by charging of traps in the silicon nitride layer upon application of a sufficiently high voltage across the gate dielectric. The induced charge results in a change of transistor turn-on voltage which can be detected as a variation of device conductance for a given gate voltage. They also noted that recent work indicated that the MNOS structure is compatible with silicon-gate as well as metal-gate MOS processing technology. Hence, full advantage can be taken of the improved performance and chip size reduction of silicon gate technology without compromising the speed and retention characteristics of the MNOS storage transistor.

The application of MNOS transistors as non-volatile storage elements in digital counters was first described by Flad et al. in 1969. The basic non-volatile flip-flop configuration is shown in Figure 20. It is a conventional cross-coupled MOS bistable circuit in which two MNOS transistors have been added in series with the set-reset transistors. The MNOS transistors have a hysteresis characteristic identical to that of the conventional memory transistors. Applications of MNOS field-effect transistors have been proposed for both logic circuits (as an alternative for stable MOS transistors) and non-volatile memory arrays. The MNOS transistor for logic circuit applications combines the low surface state density of the thermal oxide (300-500 $\overset{\circ}{\text{A}}$)-silicon interface with the passivation properties and high dielectric constant of the silicon nitride layer (500-700 $\overset{\circ}{\text{A}}$),

according to Frohman-Bentchkowsky and Forsythe (3). On the other hand, the
same basic transistor structure with a thin thermal oxide (approximately 50 Å)
has recently been shown to exhibit hysteresis behavior of turn-on voltage as
a function of applied gate voltage, resulting from charge storage at the
silicon nitride-silicon dioxide interface. The storage function associated
with the hysteresis characteristic leads to the potential application of
variable turn-on voltage MNOS transistors in random access and alterable read-
only semiconductor memories. Frohman-Bentchkowsky and Forsythe (3) noted
that both the long term stability of MNOS logic transistors and the storage
retention capability of the memory transistors are a sensitive function of
the silicon nitride current-field characteristics. The electronic charge
transport through the silicon nitride layer under an applied voltage leads
to a potential long term instability in MNOS logic transistors. However,
long term stability (ΔV_T <0.5 V after 10^5 hours at 125°C) can be acieved by
deposition of a low conductivity silicon nitride layer.

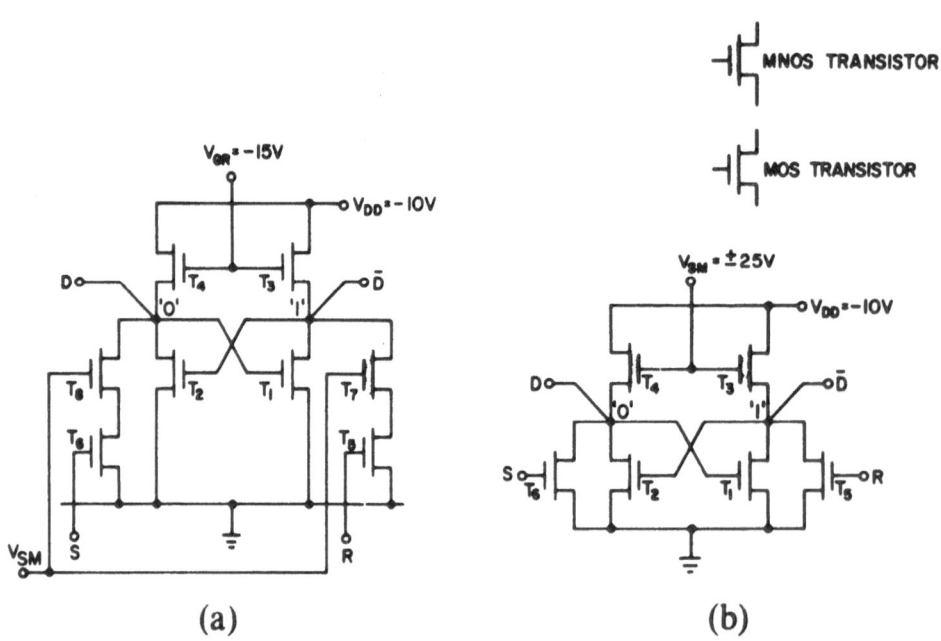

Figure 20. Nonvolatile Flip-Flop Configurations,
(a) Static and (b) Dynamic.
(Frohman-Bentchkowsky (2))

50

According to Ross et al. (1), the MNOS (metal-nitride-oxide semiconductor) memory transistor is similar in geometry and fabrication technique to standard MOS (metal-oxide-semiconductor) devices, except for the gate insulator, which is a two layer structure of silicon dioxide nearest the silicon, with silicon nitride on top of the silicon dioxide. Donor-type defect states or traps, exist at or near the interface between these two insulators and the states may be charged and discharged by the application of electric fields to the insulators. The amount of charge in the traps influences the surface potential of the silicon, and can therefore be used to alter the threshold voltage of a transistor fabricated with this double insulator structure. These investigators used silicon dioxide layers 15 to 30 $\overset{\circ}{A}$ thick, thereby employing direct tunneling between the traps and the silicon as the mechanism responsible for charge transfer. This approach combines the advantages of fast write-in at moderate voltage levels with long-term memory retentions. They also presented some results on the effect of growth of the nitride in the range of 650-1100°C upon the initial charge storage, maximum charge storage, memory retention, and bulk conductance of the insulator structure.

A great deal of effort has also been applied to the development of the MNOS memory transistor, in which charge can be stored at or near the interface between the two insulators. Investigators active in this area include Ross et al., Pao and O'Connell, Sewell et al., Wallmark and Scott, and Frohman-Bentchkowsky to mention just a few.

Silicon nitride is being actively explored at the present time as a MNOS storage transistor or NDRO (non-destructive read-out), non-volatile, single element, semiconductor memory device by Hsia and Holland, and Patterson. They have reviewed, at the recent 1970 LSI Memories Session, WESCON, the memory operating features of such a transistor with reference to its system application parameters (which include device speed and lifetime storage). They furthermore presented, design considerations and experimental data on a 256-bit silicon nitride memory array which will find applications in MSI non-volatile shift registers, LSI memories, especially in the area of electrically alterable read-only memories, buffer memories, and bulk storage. Their MNOS storage transistor had the structure shown in Figure 21.

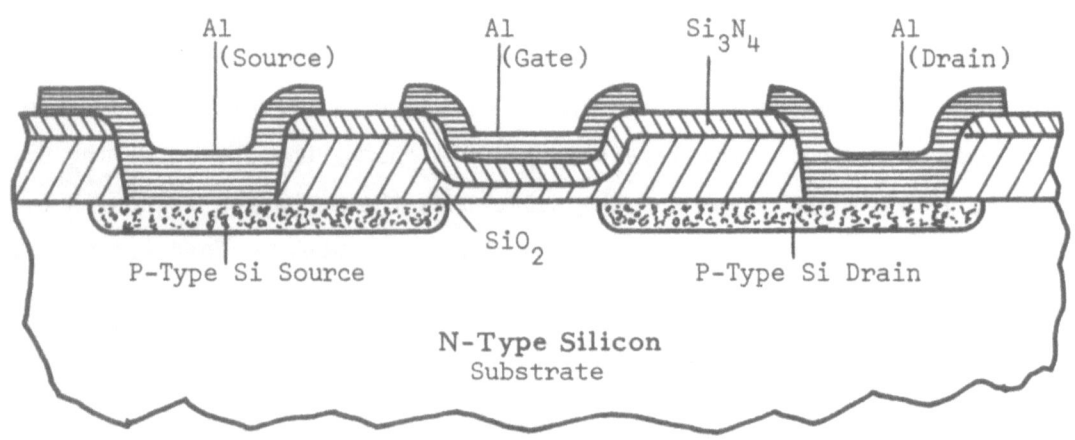

Figure 21. MNOS Storage Transistor.
(Hsia and Holland)

The silicon nitride layer is several hundred $\overset{\circ}{A}$ thick and a thin layer of SiO_2 less than 100 $\overset{\circ}{A}$ is directly over the gate channel. The nitride acts as both insulator and storage medium. Operation of the device depends upon a negative voltage exceeding a critical value applied to the MNOS gate which causes traps to be created in the SiO_2 and Si_3N_4 interface, resulting in MNOS transistor threshold change. A positive voltage of the same magnitude at the MNOS gate empties the traps and restores the transistor threshold to a lower absolute value. Data storage of over two years is clearly shown in graphic form and data storage at elevated temperatures (to 125°C) have also been evaluated. These investigators also demonstrated the feasibility of a random-access,

52

read-only memory system with a capability of performing erase and write-on of any word in the memory without disturbing the contents of other words. For a memory capacity of 2 to 4000 words by 16 to 32 bits, they estimate that the speed performance can be in the microsecond range provided bipolar circuits are used for driving and sensing. A 56-word by 8 bits feasibility model of the memory was built, using an all positive voltage system to allow for future monolithic fabrication of the bipolar circuits and to make them TTL compatible as well.

Frohman-Bentschkowsky (2) has presented a unified approach to the characterization of both the stable and variable turn-on voltage MNOS transistors. His analysis is based on an extensive investigation of charge transport and storage in MNOS structures. The devices with which he worked had a gate dielectric thin layer (15 to 500 Å) of thermally grown SiO_2 over which a layer of silicon nitride (500 to 1000 Å) was deposited. The latter was accomplished by the vapor reaction of $SiCl_4$ or SiH_4 with NH_3 over the temperature range from 800 to 1200°C. The resulting transistor had the structure shown in Figure 22.

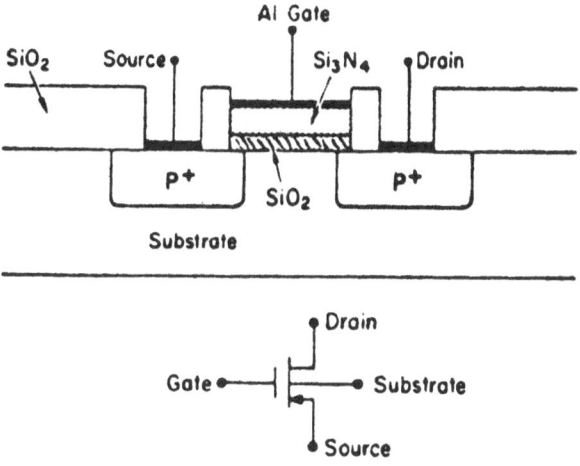

Figure 22. Cross Section of a p-Channel MNOS
Transistor with the Symbolic
Electrical Representation.
(Frohman-Bentschkowsky (2))

53

For low values of negative voltage applied to the gate, the MNOS transistor behaves like a conventional p-channel MOS field-effect transistor. On application of a sufficiently high positive voltage to the gate, electrons will tunnel from the silicon conduction band into the silicon dioxide conduction band where they will drop into traps in the silicon nitride, resulting in the accumulation of a negative charge at the dielectric interface. As pointed out in the companion EPIC VOL. 3 on SILICON NITRIDE, the exact nature of the traps is not too well understood at present. Their important feature, from the standpoint of transistor operation, is that they can be charged and discharged by an externally applied voltage. Etching experiments performed by Pao and O'Connell have shown that the accumulated charge is located in the silicon nitride within less than 100 $\overset{\circ}{A}$ of the Si_3N_4-SiO_2 interface. With this understanding of the physical mechanisms underlying transistor operation as well as optimization of transistor structure, Frohman-Bentchkowsky (2) was able to demonstrate the feasibility of fabricating an integrated 9-bit word-organized memory and a nonvolatile flip-flop. The memory cycle time was 2.0 microseconds with WRITE voltages of ±25 volts. Variable turn-on voltage MNOS transistors were used for the memory function and MNOS field-effect transistors for the driving circuitry.

Dill and coworkers (1, 2) at Hughes Aircraft Co. have recently developed a new stacked gate MNOS tetrode (4-terminal) device which had 1500 $\overset{\circ}{A}$ of thermal SiO_2 and 2000 $\overset{\circ}{A}$ of silicon nitride as a sandwich and operated on a new MNOS charge storage effect (i.e., electrons are trapped in the Si_3N_4). Applications contemplated for this device are read-only memories. The memory element (which is non-volatile and naturally suited for three-dimensional address) is bulk erasable and can be rewritten. Two-layer metallization, however, is required and solved by a Mo-Si_3N_4-Al process. The stacked gate tetrode developed as a high drain breakdown device, differs from the usual MOSFET in that the control gate is offset from the drain, leaving an offset channel which is controlled by a second stacked gate over

a thick insulator. A section through the channel region of this structure
is shown in Figure 23.

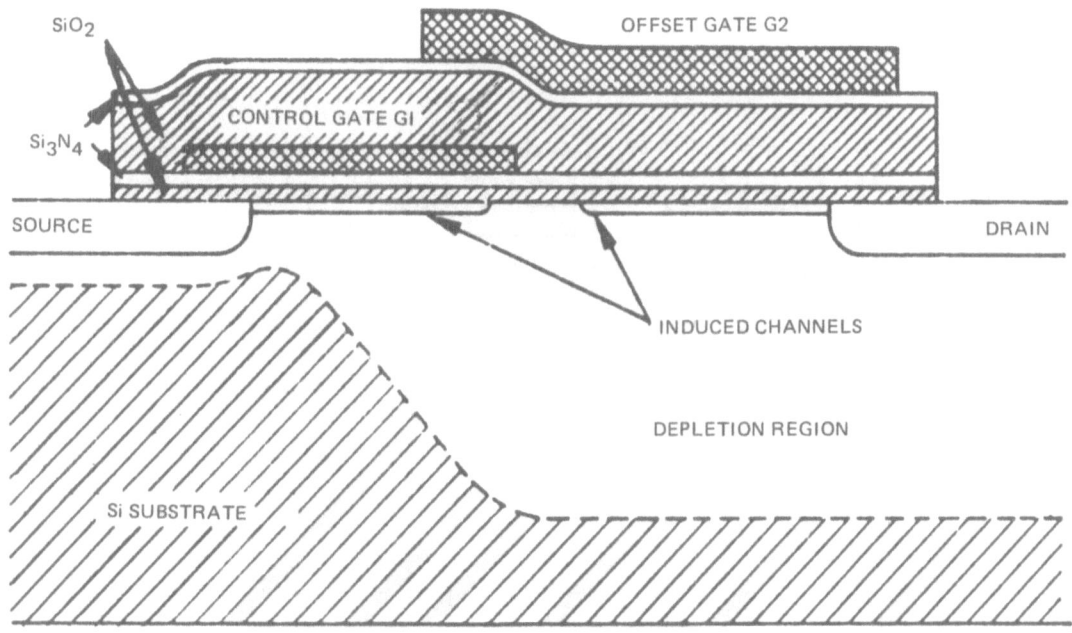

Figure 23.   Cut Through the Channel Region of the
             Stacked Gate Tetrode. (Dill and Tombs (1))

The structure consists of the source, drain, control gate, offset gate and
insulators. The depletion region and the induced channels under the control
gate and offset gate are also shown. Bias applied to the stacked gate forms
a conduction channel connecting the control gate channel to the drain. The
conductivity of this bridging channel can be adjusted so that the surface
field is reduced, thus allowing the drain voltage breakdown to approach the
bulk breakdown of the drain junction. The operation involves the injection
of carriers from the channel region under the control gate into traps in the

insulator underneath the stacked gate. The proper bias condition on the stacked gate and drain causes carriers from the pinchoff region near the control gate to be accelerated towards the insulator interface. The injected carriers are trapped in the insulator and act as charge centers which change the threshold voltage of the stacked gate. The memory cell construction utilizes the above structure but requires, in addition, two-level metallization. The read, write, and erase operational modes for the memory are tabulated in Figure 24.

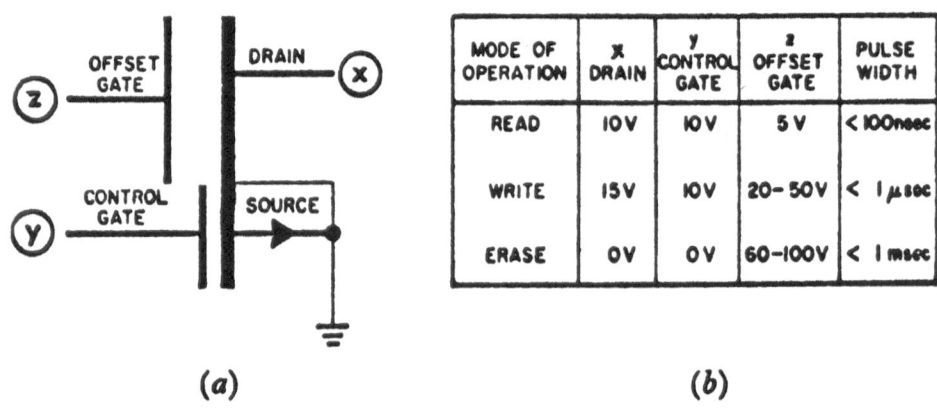

MODE OF OPERATION	x DRAIN	y CONTROL GATE	z OFFSET GATE	PULSE WIDTH
READ	10V	10V	5 V	< 100nsec
WRITE	15V	10V	20-50V	< 1 μsec
ERASE	0V	0V	60-100V	< 1 msec

(a)   *(b)*

Figure 24. (a) Injection Storage Element with
3-dimensional Selection.
(b) Read, Write and Erase Operation.
(Dill and Tombs (1))

A 16-word by 8-bit memory is described and the advantages as well as problems of such a memory element is summarized in Figure 25. p-Channel tetrodes with excellent stability have also been built.

Large memory arrays can be built with a unique memory element, utilizing silicon nitride films in conjunction with silicon dioxide and amorphous silicon gate structures. Such devices and structures were reported by Vadasz and coworkers. The structure is illustrated in Figure 26. De-

tails of the preparation of this structure are given in the paper as well
as a discussion of merits of the MOS design benefits achieved by this construction. Intel markets a 256-bit, random-access memory (RAM) element
designated as Intel-1101 based on this silicon nitride/silicon dioxide/
silicon gate structure. The chip size is 3.10 x 2.8 mm and is packaged in
a 16-lead dual inline package (DIP). An 8-bit address code will select
any one of 256 bits for either read or write operation. All address input
logic levels are compatible with standard bipolar TTL or DTL logic levels.
The mode of operation ("read" or "write") is determined by the R/W input
control. The company claims larger memory arrays can easily be built with
this memory element. Sixteen packages can make a 4096-word x 1-bit memory
by simply OR-typing the outputs, paralleling all address lines, and decoding
to each inhibit lead.

ADVANTAGES	PROBLEMS
3 Dimensional Address	Stacked Gates Require 2 Layer Metallization
Non Volatile	
Low Power Dissipation	Enhancement n-Channel Logic
Radiation Hard	
High Packing Density (Up to 1000 Bits with Decoding Circuitry on 100/100 Mil Die)	
Low Fields for Read and Write Cycle Assure High Reliability	
High Read Speed (<100 nsec)	
High Write Speed (<1μsec)	
Bulk Erasable	

Figure 25. Summary of Advantages and Problems
with the Injection Memory.
(Dill and Tombs (1))

57

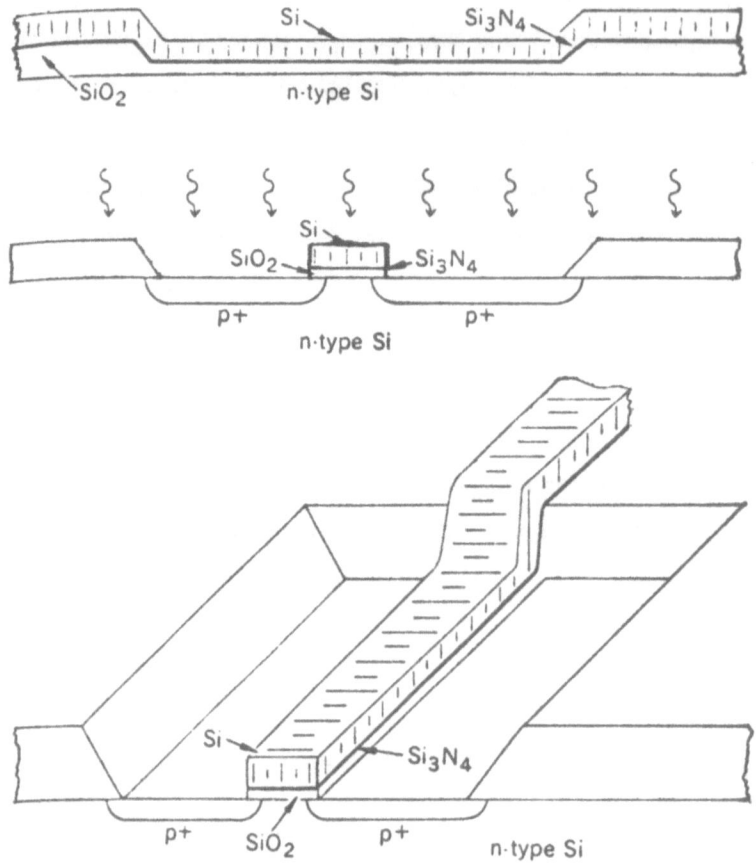

Figure 26. Large Memory Arrays Utilizing Silicon
Nitride Films with Silicon Gate Structures.
(Vadasz)

MNOS memory devices were reviewed by Wegener (2) in a recent paper. He
noted there are four simple criteria by which the competitive impact of
the MNOS memory can be assessed. The first is uniqueness; the features
that the MNOS devices possess that other devices do not. On this level,
the MNOS device excels by its non-volatility over other semiconductors.
In comparison with core and wire memories, its compatibility with other
semiconductor technologies is outstanding; in addition, it has inherently

non-destructive read-out characteristics. Finally, its high packing density results in particularly small size and low weight. In terms of operating speed, the MNOS is on par with that of the MOS memory technology. It is therefore slower than bipolar memories, and perhaps wire; but it is faster than core. The power requirements of the MNOS memory are outstandingly low. Due to its non-volatility, it uses zero standby power. This makes it equal to the magnetic technologies, and superior to the semiconductor technologies. The power dissipated during writing and reading may be as low as microwatts, if the proper circuit design is employed. In this respect, the MNOS memory may be as much as an order of magnitude better than MOS, core and wire technologies, and two orders of magnitude better than bipolar memories.

In comparing MNOS and MOS transistors, Wegener noted that they both have the same basic structure and the same electrical characteristics. They differ in one important aspect: For the MOS, the threshold voltage is fixed at a constant value. In contrast, the MNOS transistor provides the added capability that this threshold voltage can be set and reset electrically at predetermined high (-12V) and low (-2V) values. Once set, a specific threshold voltage will remain near its value for an extended and predictable length of time, even after any source of power to the device has been removed. Writing is accomplished by the application of a relatively high (25-30V) voltage between the substrate of the transistor and the gate. Voltages significantly lower, such as the interrogation voltage (-7V) applied for read-out, have a negligible effect on the prevailing threshold voltage. The amount of threshold voltage change depends on the polarity of the applied voltage.

The mechanism of the threshold voltage change is well understood according to Wegener. It depends on the accumulation or the removal of charges at the interface of two regions differing in conductivity when a potential is applied across both. These two regions are provided by the silicon oxide and silicon nitride layers forming the gate of the transistor from which

59

the device derives one of its names. The other important feature is a highly nonlinear conduction in the more conductive silicon nitride layer. Writing is done at high voltages, leading to very high currents, and therefore a very high rate of charge accumulation in the gate dielectric. Reading and storing is done at low voltages, resulting in only very low currents, and therefore a very low rate of charge decay. Both oxide and nitride can provide the nonlinearity and the magnitude of current required for a useful memory element. Memory elements with widely differing writing and storage properties can be made by the judicious structuring of the transistor gate with these materials. Devices with predictable writing delays ranging from 10^{-7} to 1 second have been made. These values have opened up a wide range of memory applications for MNOS devices. They include electrically programmable read-only memories having MNOS structures with long storage times. They also include random-access memories with MNOS structures having fast writing times.

Flad et al. described the application of MNOS transistors in a preset counter with nonvolatile memory. Such transistors (p-channel field-effect devices) exhibit a resettable threshold voltage. Their basic MNOS structure comprised two p-type regions which are diffused into an n-type silicon substrate. The gate insulator consisted of a layer of silicon nitride (several hundred $\overset{\circ}{A}$ thick) over a layer of silicon dioxide (approximately 100 $\overset{\circ}{A}$ thick). As noted above, this device has characteristics which are typical of conventional enhancement type MOS transistors when operated at normal gate voltage levels. However, when the positive voltage applied between gate and substrate exceeds a critical value which is determined by the device design (typically +50V), a transition from enhancement to depletion type characteristics occurs during the gate polarization interval. The resulting depletion-type characteristic represents a normally-on conduction state (threshold voltage of approximately +2V) when the polarization voltage is removed. Subsequently, when a negative-gate polarization voltage of similar magnitude is applied, the opposite transition occurs and the device characteristics shift to the initial normally-off state ($V_{th}\sim3V$). The two, stable, threshold voltage levels which result

60

when the gate is alternately polarized with the critical voltage constitutes the non-volatile memory effect which is utilized in this circuit. These investigators found that using MNOS transistors throughout the circuit makes it possible to integrate counter and memory on a single chip. In addition, the MNOS load devices may be preset to the depletion mode, thereby eliminating the threshold voltage drop, characteristic of common drain-gate configuration. Other design features of the circuit are a source to drain breakdown voltage independent of the required polarization voltage and a steering circuit which eliminates undesirable gate charging effects.

MNOS memory transistors were also fabricated at RCA Laboratories by Goodman and coworkers and comprised a SiO_2 layer (16-35 Å) for operating in the direct-tunneling mode plus a Si_3N_4 layer deposited in the range from 650 to 1100°C using silane and ammonia (1:10,000 respectively). The optimum SiO_2 thickness was found to be 20 Å, the nitride deposition temperature 700°C. The nitride film characteristics were evaluated by using the MNOS capacitor structure and compared in a series of graphs as a function of deposition temperature. There is a very definite relationship between the deposition temperature and the amount and type of charge as well as the charge storage behavior as a function of time. Their experimental data with respect to charges stored are shown in Figure 27 and Figures 36 and 37 (see Ross et al. page 70).

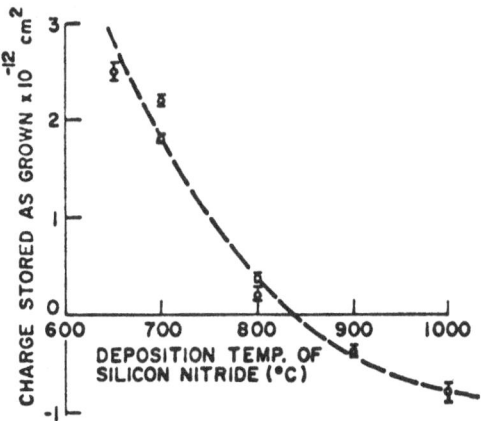

Figure 27. Initial Charge Stored in MNOS Structure
Versus Deposition Temperature of
Silicon Nitride. (Goodman et al.)

61

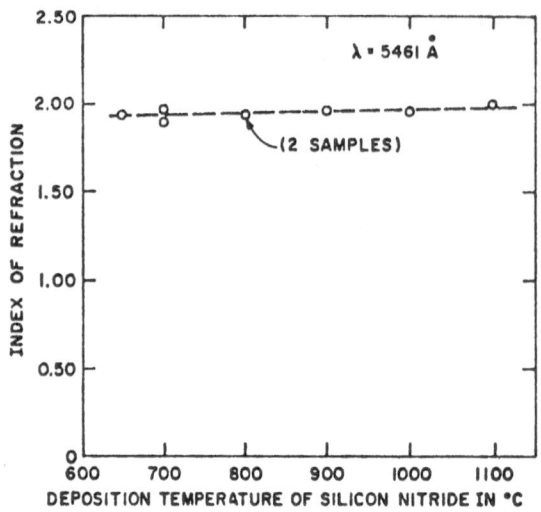

Figure 28.   Index of Refraction of Silicon
Nitride as a Function of
Deposition Temperature.

(Goodman et al.)

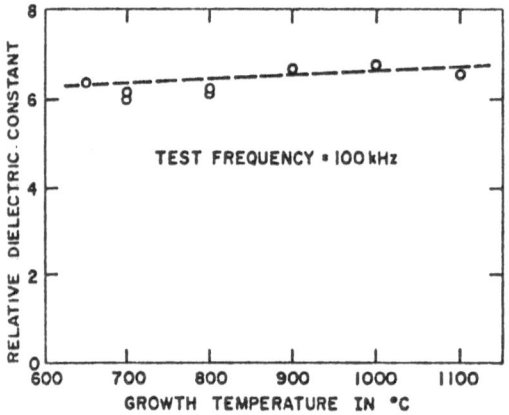

Figure 29.   Relative Dielectric Constant of
Silicon Nitride as a Function
of Deposition Temperature.

(Goodman et al.)

As seen in Figures 28 and 29, the refractive index and dielectric constant
properties showed virtually no variation over the deposition temperature
range 650 to 1100°C. It is quite significant to note the variation of initial charge stored in the MNOS structure from positive to negative in this

temperature regime. Also of interest is the maximum positive stored charge for negative potential which results in the parabolic shape shown above. The maximum positive charge stored is also quite dependent upon the SiO_2 thickness as shown in Figure 30.

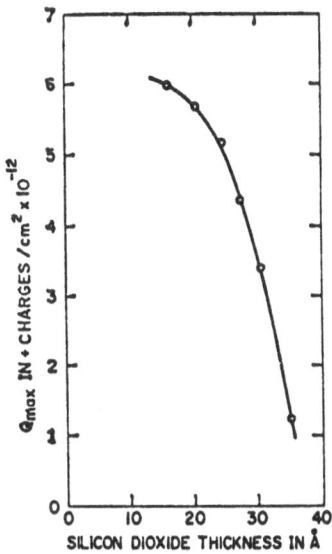

Figure 30. Maximum Charge Stored
Versus SiO2 Thickness.
(Goodman et al.)

Their explanation for the trap behavior is the following: Traps exist at or near the interface between the SiO_2 and the Si_3N_4. These traps can be charged and discharged by the application of a sufficiently large voltage of suitable polarity to the gate electrode. The charge states of the traps influence the silicon surface potential and, therefore, can affect the threshold voltage of the MNOS transistor. Each trap is assumed to be electrically neutral when filled with an electron and positively charged when empty. Application of a large negative voltage to the gate raises the traps in energy relative to the silicon, permitting the traps that are energetically above the conduction band edge of the silicon, and are within a probable tunneling distance, to give up an electron to the silicon. When the gate voltage is returned to ground, the empty traps, which are now positively charged, are energetically opposite the forbidden gap of the silicon, and are therefore metastable. Application of a large positive voltage permits electrons to tunnel from the silicon valence band back to the traps, thereby neutralizing their positive charge and returning them to their original charge state. The energy band diagrams of these various bias conditions are depicted in Figure 31.

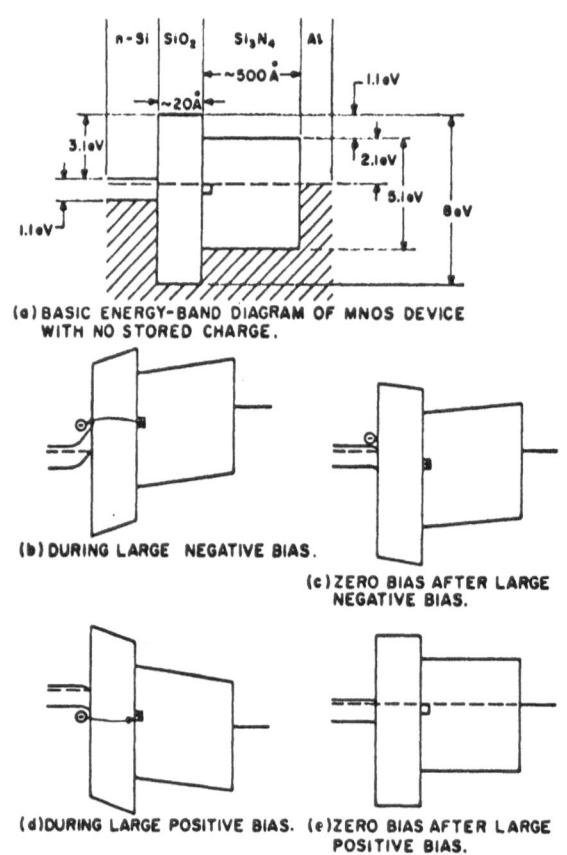

(a) BASIC ENERGY-BAND DIAGRAM OF MNOS DEVICE
WITH NO STORED CHARGE.

(b) DURING LARGE NEGATIVE BIAS.

(c) ZERO BIAS AFTER LARGE
NEGATIVE BIAS.

(d) DURING LARGE POSITIVE BIAS.   (e) ZERO BIAS AFTER LARGE
POSITIVE BIAS.

Figure 31.   Energy-Band Diagrams of an MNOS
Device for Various Bias Conditions.
(Goodman et al.)

Utilizing the above charge storage/trapping explanations and deposition/ layer approach, memory transistors were fabricated using the optimum values cited above. Gate threshold voltage shifts of 2 volts were obtained using switching pulses of 100 nanoseconds duration and ±34 volts amplitude.

Another band picture of the Al-SiO$_2$-Si$_3$N$_4$-Si structure has been proposed by Wallmark and Scott and used as a physical model to explain the memory effects or mechanisms involved when this structure is used as a memory transistor. The energy band picture was assembled from energy level data of several other investigators as shown in Figure 32.

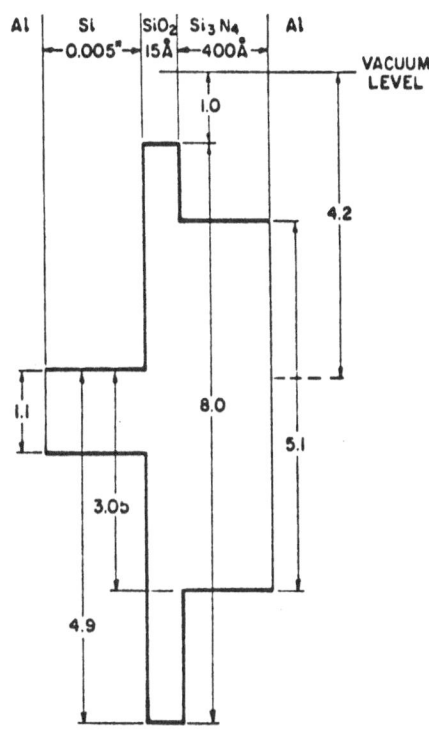

Figure 32. Band Diagram of the Silicon-Silicon Dioxide-
Silicon Nitride-Aluminum System with Quanti-
tative Values of the Energy Levels.
(Wallmark and Scott)

Sewell and coworkers (2) at Sperry Rand Research Center have also developed a simple two-layer model for charge storage in the metal-nitride-oxide-semiconductor structure (MI$_2$I$_1$-S) or device in which the times for charging and discharging are expressed in closed-form expressions depending on the conduction properties, the thicknesses, and the dielectric constants of the two layers. Their data is shown in Figures 33 and 34 as evidence of the conformance to the model.

65

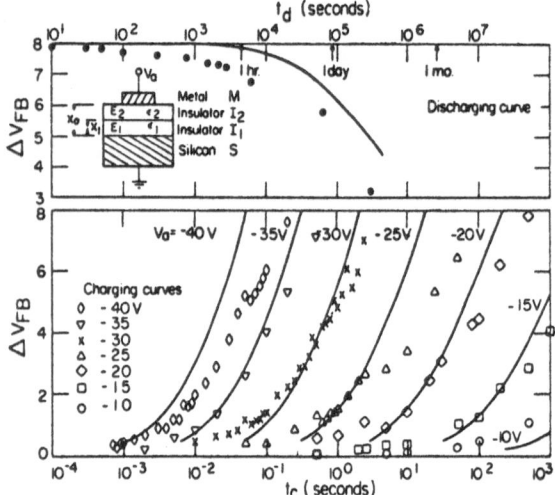

Figure 33. Comparison of experimental data with calculated curves showing the effect of $-V_a$ as a parameter. The values of $x_O= 1200$ Å, $x_1= 350$ Å, and the nitride conductivity were all obtained experimentally from the sample for which the charging measurements were taken. Decay data from the same sample are also shown. Insert shows schematic of the MI_2I_1S device. (Sewell Jr. et al.)

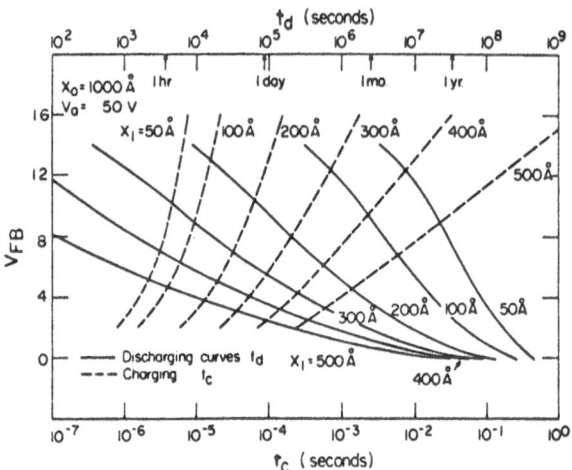

Figure 34. Calculated curves showing the effect of oxide thickness on the charging and discharging times. Here $x_O= 100$ Å and $V_a= 50$ V were arbitrarily chosen, while experimental values of $\alpha= 1.588 \times 10^{-2}$ (cm/V)$^{1/2}$ and $J_O= 1.363 \times 10^{-18}$ A/cm^2 were used. Clearly the optimum memory characteristics are obtained for small values of x_1, although for quite small x_1 the conduction mechanism of the oxide may change drastically.
(Sewell Jr. et al.)

Since 1967, Lewis, Sewell, and Wegener (3,4) at Sperry Rand have investigated MI_2I_1S and MIS capacitor memory devices that have both adaptive and memory characteristics in order to obtain a better understanding of this type of device, attempting to attain a thorough knowledge of all the previously-mentioned participating conduction mechanisms to achieve device predictability. Silicon wafers were subjected to high temperature surface treatments (hydrogen and ammonia ambients at 1250°C) prior to the deposition of silicon nitride (which was performed in hydrogen at 900°C). The effect of light on the silicon nitride charge storage properties was shown to be related to a space charge layer in the silicon adjacent to the insulator. With a relatively intense light directed on the surface of the device, this space charge layer can be practically eliminated.

They had also noticed during the wafer probing of memory capacitors, that a significant enhancement in charge storage occurred when the light from a microscope light was directed onto the surface of the wafer concurrently with the application of a negative charging pulse on the gate electrode. There was no enhancement when a positive charging pulse was applied. This effect was observed on all memory structures that were fabricated on n-type silicon wafers. The current vs negative gate voltage characteristics of

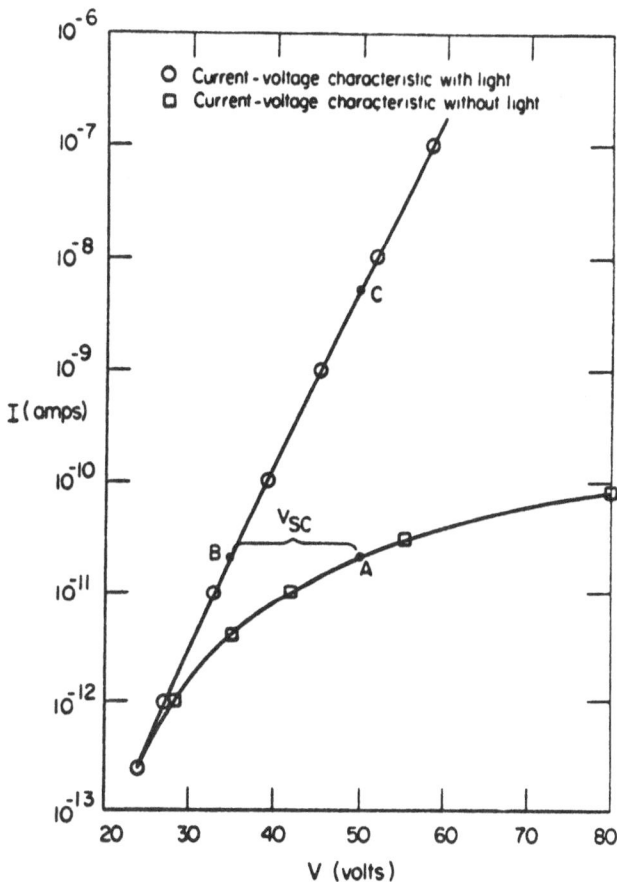

Figure 35. Current vs Negative Gate Voltage Characteristics of a Memory
Capacitor with and without Illumination. (Lewis & Wegener (3))

a memory capacitor is shown both with and without light directed on the surface of the device in Figure 35. With light on the device, the generation of electron-hole pairs in the silicon is great enough to cause a collapse of the space charge layer adjacent to the insulator. This space charge layer results from the depletion of electrons from the surface of the silicon when a negative voltage is applied to the gate electrode. This depletion continues as the bias voltage becomes more negative because the rate of charge flow through the silicon nitride exceeds the generation-recombination rate of carriers in the silicon space charge region.

In analyzing their various experimental results, the Sperry Rand investigators found that these memory device structures (MIS and MI_2I_1S) store charge as a result of the presence of two different conduction mechanisms. One of these mechanisms is inherent to the silicon nitride, i.e., Poole-Frenkel, and the other is a silicon-silicon nitride interface controlled mechanism. The conduction characteristic so derived was found to correspond to a Fowler-Nordheim tunneling equation. During a charging pulse, this conduction mechanism transfers charges at a greater rate than that of the bulk silicon nitride. This results in a charge accumulation in the region that is defined by the boundary between the two conduction mechanisms. Sperry's Lewis et al (4) discuss in great detail the con-

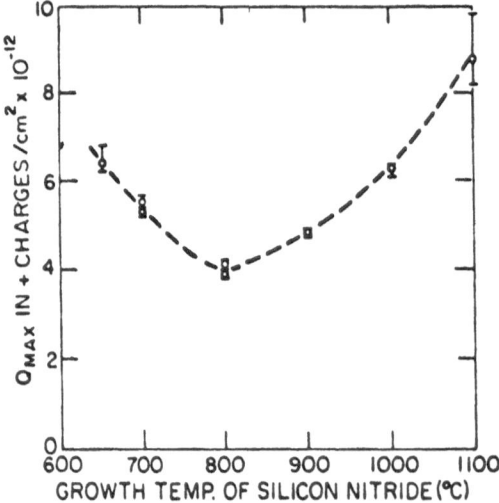

Figure 36. Maximum Positive Stored Charge for Negative
Potential vs. Growth Temperature of Silicon Nitride.

(Ross et al. (2))

68

duction mechanisms in the insulator structure of the MI_2I_1S memory device including equations necessary to describe the transient behavior of the device.

Air Force needs for an electrically alterable nonvolatile read-only memory has resulted in various contract studies to develop memories of this type. As discussed before, such a memory element is readily achieved through the physical properties of thin insulating films, in particular, silicon nitride and silicon dioxide. These films when fabricated into the gate of a field-effect transistor provide an electrically alterable threshold voltage, and therefore digital storage when the memory element is used in a suitable array. Using this MNOS variable threshold FET as a basic memory element or cell, Sperry Rand Research Center (under Air Force contract no. F33615-69-C-1290) has been developing a 1024-bit monolithic array capable of performing an electrically alterable read-only memory function, which is nonvolatile with a nondestructive read-out capability.

Ross et al. (2) at RCA Laboratories reported experimental data on the effect of varying the Si_3N_4 growth temperature on the charge storage characteristics of the MNOS structure. Of primary importance is the maximum amount of charge which can be stored as well as the long term retention of the stored charge. Figures 36 and 37 summarize their experimental investigations.

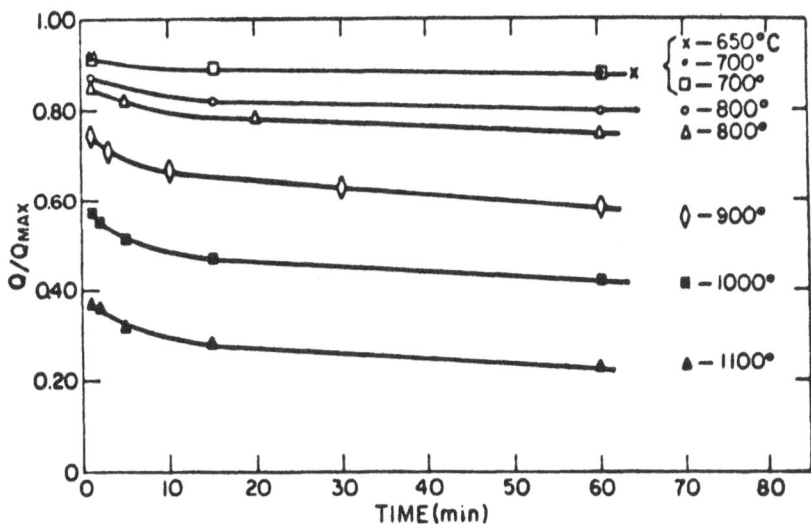

Figure 37. Charge Stored vs. Time with Growth
Temperature as a Parameter.

(Ross et al. (2))

It is readily apparent that the charge storage characteristics of the MNOS structure is dependent upon the preparation conditions of the silicon nitride. In a more recent paper, Ross and Wallmark (3) discuss the available theories to explain the switching behavior of MIS memory transistors and presented their own theory in which direct tunneling is the principal mechanism of charge transfer. The theory predicts that charge transfer in such a device is logarithmically dependent on the applied gate pulse duration and exponentially dependent on the amplitude of the applied gate pulse. The theory is shown to be in good agreement with experimental results. Ross and Wallmark (2) assumed the current mechanism to be direct tunneling from the silicon into the interface states located at the interface between the two insulators (15-25 $\overset{\circ}{A}$ SiO_2).

Ross and coworkers (4) at RCA made an in-depth study of the operational dependence of the direct-tunneling mode MNOS memory transistor on the SiO_2 and Si_3N_4 layer thicknesses and found it to depend quite critically on this parameter. In terms of the threshold separation obtained for a given applied gate voltage and a fixed thickness of Si_3N_4, a thickness of 20 $\overset{\circ}{A}$ for the SiO_2 layer is near optimum. Their studies also showed that the operating characteristics were relatively insensitive to the value of the ambient temperature in the range from room temperature to 125°C. Additionally, as the SiO_2 thickness is increased, the rate of loss of stored charge decreases. Figure 38 shows the structure of their MNOS memory transistor.

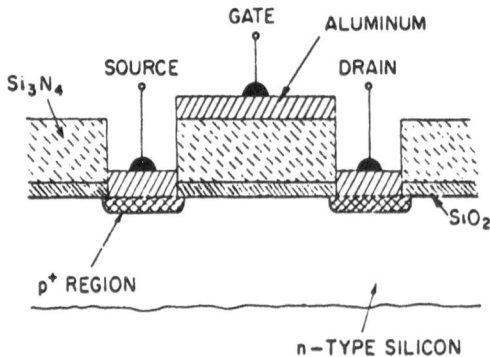

Figure 38. Schematic Diagram of a MNOS Memory
Transistor. (Ross et al. (4))

The transistor is a standard insulated-gate, field-effect transistor, except for the gate insulator, which is a double-layer structure of silicon nitride and silicon dioxide. This MNOS structure as shown previously (or discussed previously) exhibits charge storage effects that are associated with traps at or near the interface between the SiO_2 and the Si_3N_4. The amount and sign of the charge stored in the traps affects the surface potential of the silicon. The charge state of the traps and, therefore, the threshold voltage of the MNOS transistor, is electronically alterable by the application of large gate fields. Typical transfer characteristics for a direct-tunneling mode MNOS memory transistor is illustrated in Figure 39.

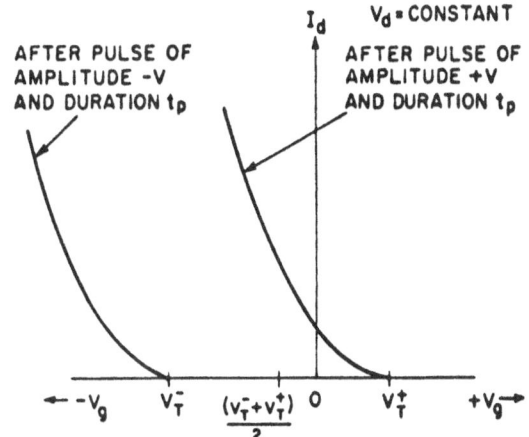

Figure 39. Illustration of Typical Transfer Characteristics for a Direct-Tunneling Mode MNOS Memory Transistor.

(Ross et al. (4))

The amount of threshold voltage shift that results from the application of a given gate voltage is a function of the thickness of both the SiO_2 and Si_3N_4 layers. Figure 40 depicts the effect of varying the silicon nitride layer while holding the SiO_2 Constant at 20 Å. Also important to the utilization of the MNOS transistor in practical applications are the charge-retention characteristics, once a threshold condition has been selected. Charge loss experiments showed that charge retention was 85%, 5 weeks after

71

storage and 80% ten years after storage. Figure 41 demonstrates that the charge loss decreases with increasing SiO_2 thickness. These results support the concept that as the SiO_2 thickness increases, the tunneling barrier width between the traps and the silicon is increased and the tunneling probability, therefore, decreases.

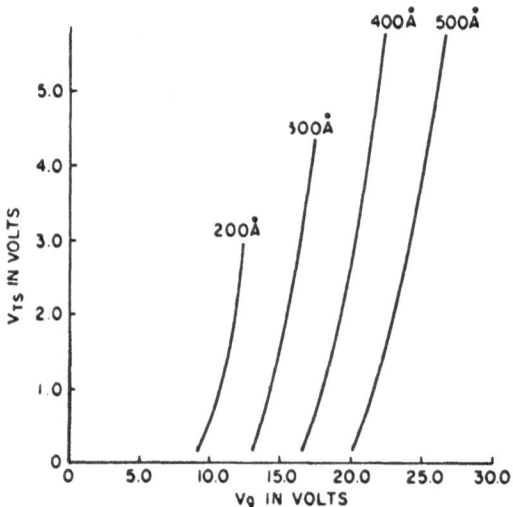

Figure 40. V_{ts} Versus V_g for a Silicon Dioxide Thickness of 20 Å, a Pulse Duration of 1 x 10^{-3} Second and the Silicon Nitride Scaled to Various Thicknesses. (Ross et al. (4))

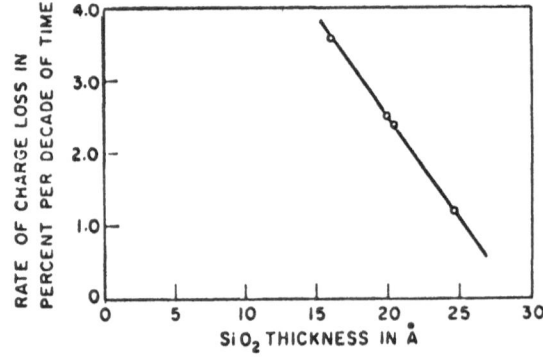

Figure 41. Rate of Loss of Stored Charge Versus Thickness of Silicon Dioxide. (Ross et al. (4))

Charge-storage effects in MNOS structures were also investigated by McDonald (Fairchild Camera and Instrument Corp.) with regard to application in programmable read-only semiconductor memories utilizing the insulated-gate field-effect transistor as the active circuit component. By means of the stored charge in an MNOS gate over the emitter-base junction of a bipolar transistor, the base surface beneath the gate can be stably shifted from accumulation to inversion with the inversion layer beneath the gate of a suitably designed transistor acting as an efficient extension of the emitter junction. The principal advantage of this type of device over conventional MNOS devices lies in its inherent higher transconductance.

Wallmark and Scott of RCA, constructed memory transistors in the form of MIS field-effect transistors in which the gate insulator consisted of a double layer of silicon dioxide (15 $\overset{\circ}{A}$ thick) and silicon nitride (600 $\overset{\circ}{A}$ thick). They measured switching times of 0.05 to 1 microsecond and found the memory retention is more than 2000 hours at room temperature. Further, they found that the hysteresis in the gate capacitance versus gate voltage curve is caused by the charge storage in states located at the SiO_2-S_3N_4 interface. The transport of charge into and out of these states takes place by tunneling as a result of a large positive or negative gate voltage. Tunneling into and out of states in the insulator also takes place in SiO_2 which gives rise to the so-called "slow-trapping instability" which is undesirable in memory transistors. Their studies also involved correlation of flatband voltage with pulse width and duration. As the pulses became shorter, their amplitude must be increased to shift the flatband voltage curves a given amount. A very short pulse does not allow enough time for tunneling to traps far from the interface.

MNOS read-mostly memories have been developed by the Caswell Research Laboratoratories of Plessey Ltd. in England. According to an announcement, the gate lies on top of a silicon nitride layer above the silicon dioxide layer over the channel. The charge is stored at the interface between the dielectrics, in deep traps that form when the nitride is deposited on the oxide. To induce it, a voltage above a certain critical level has to be applied to the gate, and to discharge it, a similar voltage of opposite polarity is used. The charge stored can be preset by gate voltage and duration of application and it determines the threshold voltage at which the transistor turns on--which may be at any voltage over a wide spread. The transistor will retain the selected threshold when the gate voltage is removed. The use of two voltage levels a few volts apart creates a two-state storage device which can be read by applying a gate voltage between the two thresholds to detect whether the transistor is on or off; cleared by applying an opposite voltage; and written into with gate voltages above the critical level. Transistors made by the Plessey team show that the charge storage level remains much the same over several months at least.

There is still a problem in that the switching speed is not constant over the range of possible threshold levels, so that the choice of logic levels is dependent on speed considerations, and vice versa. In a read/write application, the two thresholds must have a switching time equal in both directions. Their present transistors can only permit a switching time of 10 microseconds or more, which is totally uncompetitive with established read/write stores.

These investigators note that apart from the amplitude, and duration of the switching pulse, the performance of the transistor is critically affected by the thickness of the oxide and nitride. Since the stored charge affects the threshold voltage by influencing the state of the silicon surface under the oxide, the oxide has to be as thin as possible so as to get the fastest response out of the minimum charge and lowest applied gate voltage. They reported an oxide thickness value of between 10 and 20 Å (controlled to ± 1 Å) and nitride thickness from 500 to 600 Å. Devices with a 12-A oxide switch from the low threshold of 4 volts or so to the high threshold of 8 volts, using a switching pulse of 1 microsecond and -40 volts. The clear time is 30 to 40 microseconds. The read voltage used is about 6 volts.

Other British investigators, Barraclough and coworkers (Royal Radar Establishment) have described their investigations on MIS memory transistors (p-channel type) having typically 15 to 40 Å of SiO_2 (produced by oxidation at 600°C) plus a layer of silicon nitride (typically 400 to 1000 Å) made by the vapor phase reaction of SiH_4 with NH_3 at 850°C. They observed a retention of charge with zero applied voltage for up to one month at room temperature on MNOS capacitors. More precise measurements show a small decrease (approximately 10%) in charge in the first few minutes and thereafter very little change.

The development of a MONOS memory element by adding another layer of oxide on the nitride of the conventional MNOS structure was reported by Keshavan and Lin (1,2) of Westinghouse Electric Corp. (Integrated Circuit Div.). They were hoping to prevent tunneling from the gate electrode into the traps in silicon nitride. These investigators found that a negative polarizing voltage -40 to -50 volts changes the device from enhancement mode to depletion mode, and a positive voltage from +40 to +50 volts converts it to enhancement mode again. This change of mode is reversible and the operation is comparable to that of normal MNOS memories.

CAPACITORS

Two types of capacitors or structures with silicon nitride as the dielectric medium are possible: The MNS (also called MIS) and the MNM (or MIM). The difference lies in the electrode materials. Both types have been explored by various investigators. Capacitor data (C-V) are often generated during the evaluation of other devices in order to obtain some understanding of charge storage, trapping, transport mechanisms. This section will be concerned only with data where the end result is to develop a capacitor device and not evaluate a capacitor structure for charge effects studies. C-V data for films are compiled in volume 3 of this series.

Barnes and Geesner as early as 1960 reported on the efforts of the Air Force Electronics Technology Laboratory to develop thin dielectric film silicon nitride for capacitors capable of operating satisfactorily up to and above 600°C. Thin adherent nonporous films of supposed silicon nitride were deposited pyrolytically from the vapor phase on hot molybdenum substrates. Such films, when incorporated between molybdenum plates to form capacitors, were found to maintain satisfactory dielectric properties up to and above 600°C. The films were smooth, clear, transparent, colorless, very adherent, free of pores, and suitable for capacitor construction. The best quality dielectric films were those which required at least one hour to lay down a thickness of 0.4 to 0.6 mil. This led to several patents in this area. Subsequently, they discovered that the films consisted of silicon dioxide with less than 0.3% nitrogen (attributed to small amounts of water vapor present).

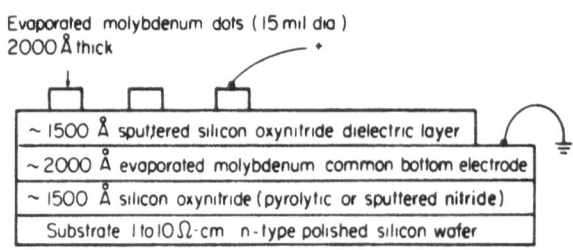

Figure 42. Schematic of MIM
Structure with a Common Bottom
Electrode. (Frank and Mosberg)

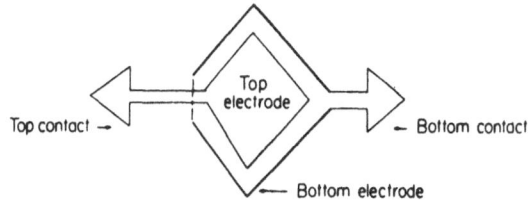

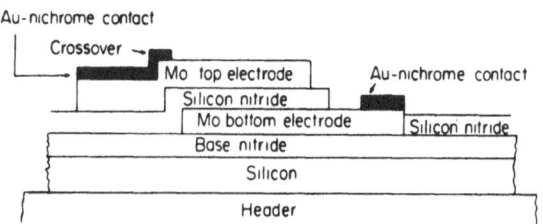

Figure 43. MIM Capacitor Structure
Used for Devices to be Mounted on
Headers.   (Frank and Mosberg)

Frank and Moberg studied an interesting experimental capacitor structure
utilizing rf sputtered silicon oxynitride as the dielectric and molybdenum
electrodes as shown in Figures 42 and 43.

In 1965 Chaikin at the Stanford Research Institute made some dielectric
property measurements on silicon nitride films as part of a thin film dielectric capacitor study.

The literature on aluminum-silicon nitride-silicon (MNS) capacitors was
reviewed by Hutchins (1), North Carolina State University, in a doctoral
thesis. His conclusions were that the deposition or preparation parameters
were not thoroughly understood at that time (1968) in view of the variations
in the electrical properties inherent in the capacitors used in his studies
(capacitors were obtained from Texas Instruments Co.). The capacitors
were fabricated on n-type silicon (10 ohm-cm) substrates using thin films
of silicon nitride (from pyrolytic deposition of SiH_4 and NH_3 in an excess
H_2 atmosphere) and filament evaporated aluminum. Further, the surface
preparation of the silicon for optimum compatibility with silicon nitride,
the deposition parameters, including the effects of impurities in the gases,

and the choice of metal were broad topics not completely understood at that time.

Savage and coworkers constructed metal-insulator-silicon (MIS) capacitors to observe the variation in the interface charge as the silicon nitride deposition conditions were altered. Initially, films were deposited at a fixed substrate temperature of 600°C while the gas ratios and rf conditions were altered. $Al-Si_3N_4-Si$ capacitors gave inconsistent results, direct currents of up to 1 microampere being obtained. These leakage currents were thought to be due to faults in the film caused by dirt on the silicon surface; a number of different cleaning procedures were tried with no significant improvement.

$M-Si_3N_4-Si$ (or MIS) capacitors using SiH_4/NH_3 gases and evaporated nickel gates were constructed by Zaininger. These were submitted to high-energy electron bombardment (125 keV and 1 MeV) to determine and study the radiation sensitivity on initial surface state density, gate thickness, oxidation procedure, surface orientation as well as annealing behavior. The main effects resulted in the generation of electron-hole pairs and the establishment of a positive space charge in the insulator by predominantly capturing the radiation-generated holes into stationary traps. Bias during the bombardment influenced the spatial distribution of the trapped holes.

The first to report on the memory behavior of a MNS capacitor were Pao and O'Connell in 1968. They found that when a MNS capacitor is subjected to single or multiple pulses, a memory behavior may be consistently demonstrated, a feature which has not been observed in the corresponding MOS structures. The essence of this behavior is the ability of the C-V curve of the MNS capacitor to return exactly to the same end limits on the voltage axis after it is given the same plus-minus pulses.

Goetzberger and Nicollian have investigated MIS capacitors which were fabricated by depositing a 1000 Å film of silicon nitride on gallium arsenide substrates. Breakdown fields and voltage studies were of major interest as related to avalanche conditions and were evaluated by C-V measurements as shown in Figure 44.

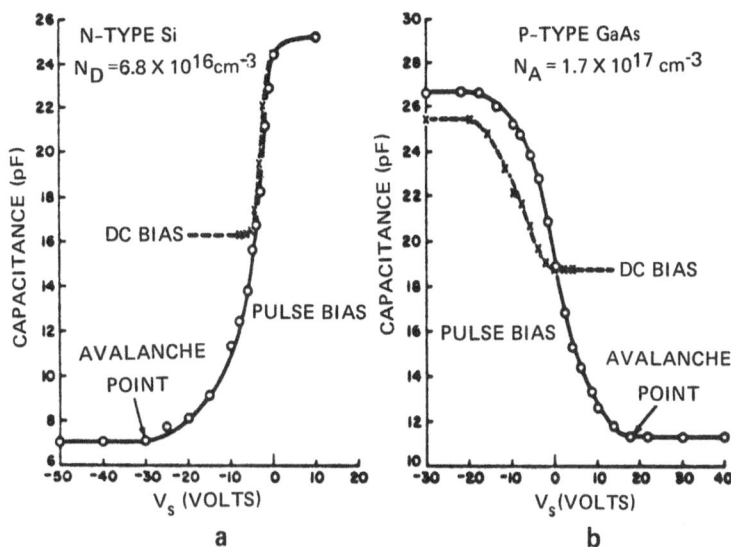

Figure 44. Curves for p-Type GaAs with 1000 Å of Deposited Si₃N₄.

The dc curve differs from the ac curve because of a slow drift effect that distorts the dc curve. (Goetzberger and Nicollian)

Wegener (1) in reporting on new concepts of adaptive devices which resulted in the development of a MNS-VTT (insulated gate field-effect transistor device possessing an electrically alterable threshold voltage), has presented useful MNS capacitor storage data and information. The MNS structure of their device is illustrated in Figure 45.

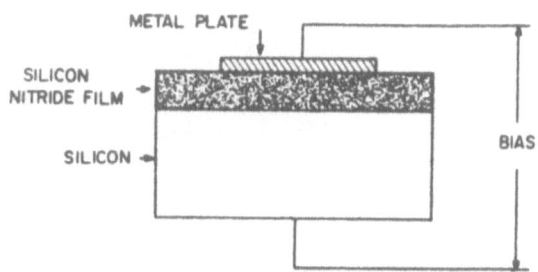

Figure 45. MNS Structure. (Wegener (1))

A qualitative picture of MNS capacitor behavior involves the problem of how the capacitance varies with the bias applied to the metal electrode and involves the behavior of the majority and minority carriers. For example, if a large negative bias is applied to the metal field plate, holes will be attracted to the silicon (e.g. p-type Si) surface, which, with the large accumulation of majority carriers, will then behave much like a metal. The capacitance measured is that of the insulator layer alone. Now, if a small positive bias is applied, holes are repelled, forming a region at the silicon surface that is depleted of majority carriers. The effective width of the dielectric will now be increased by this depletion region, and the measured capacitance will begin to decrease. A further increase in positive bias will further deplete the surface region and cause an additional decrease in capacitance. However, at some larger positive bias, there will be an appreciable accumulation of minority carriers at the interface. At this point, the depletion region width will approach a maximum, and electrons will begin to form an inversion region at the interface. If the lifetime of the minority carriers is such that they can follow the applied frequency, the capacitance will rise, eventually approaching the original capacitance due to the insulator only. If the minority carriers cannot follow the frequency of the applied signal, then the capacitance will approach a minimum value and finally become independent of voltage as illustrated in Figure 46. The pattern of behavior is for ideal structures but Wegener [1] found that when MNS capacitors undergo special processing, the successive traces on a C-V plot showed the curve displaced from immediately preceding ones. Holding the voltage on the capacitor at either the positive or the negative extremes of the voltage ramp for some time resulted in maximum displacement of the C-V plot. By further investigation and control of the process technology, they were able to produce MNS structures that displayed controllable amounts of hysteresis.

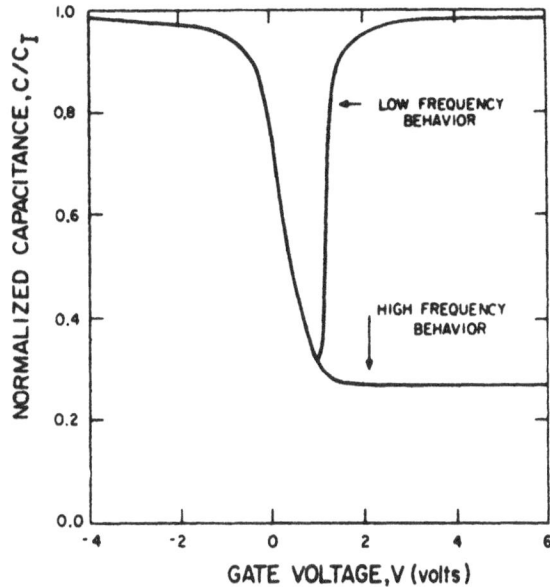

Figure 46. Theoretical C-V Curves. (Wegener (1))

The Sperry Rand researchers found the C-V hysteresis plots can be divided into two types: In the first category are MNS structures which show shifts in characteristics which are monotonically related to the difference between the applied voltage and the inversion voltage present in the device. In the second type of structure, shifts in the C-V plot only occur when some threshold level of voltage has been exceeded. Additional studies of the MNS capacitor storage phenomena by varying pulse durations, voltage amplitudes, positive and negative pulses and their effect on the "inversion voltage", led to the development of a MNS variable threshold voltage transistor (MNS-VTT) which is described in another section.

Elliott and coworkers at the University of Salford (Great Britain) fabricated $Al-Si_3N_4-Al$ and $Al-Si_3N_4-Au$ devices using electron beam evaporation for the silicon nitride at pressures of $10^{-5}-10^{-6}$ Torr along with heated substrates. If the substrates were maintained below 200°C the nitride films were found to be mechanically unstable and it proved difficult to produce a useful device. With the substrate above 200°C, the resulting dielectric films were found to be stable and transparent, although the failure rate was high. Reproducible results were obtained from a number of different devices. I-V characteristics for a number of devices were measured and a typical plot is shown in Figure 47 for an Al-Si$_3$N$_4$-Al capacitor (nitride thickness = 1650 Å).

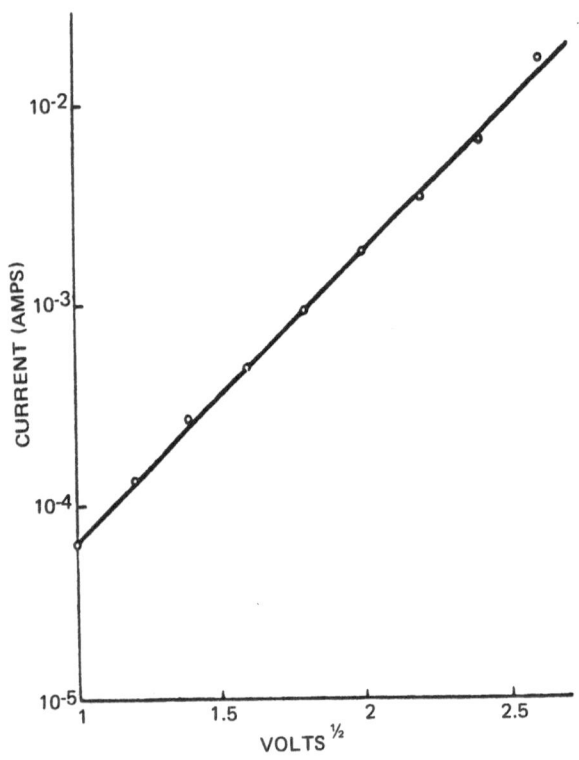

Figure 47.   Typical Current-Voltage Relationship
             for an Evaporated Si$_3$N$_4$ Film 1650 Å Thick.
             (Elliott)

An Al-Si$_3$N$_4$-Au device produced by the same technique (electron beam evaporation) gave conduction effects which were independent of electrode polarity, and it seems likely that a Poole-Frenkel type of conduction mechanism is operating in this case.

Deal et al. (2) have reported on and compared typical capacitance-voltage plots for silicon nitride MIS structures as shown in Figure 48.

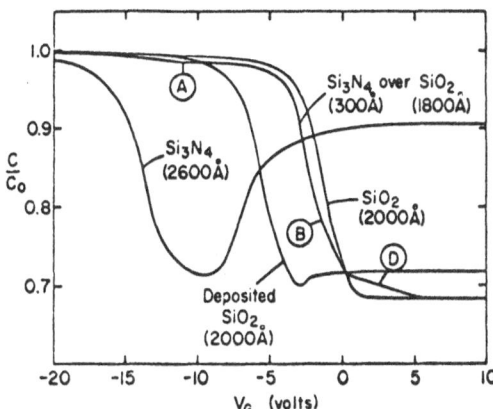

Figure 48. Typical Capacitance-Voltage Plots
for MIS Structures (f= 100 KHz)
(Deal et al. (2))

Hu (1) was the first to report an instability related to silicon nitride films on silicon. He noted that a displacement along the voltage axis of MIS capacitance-voltage plots occurred at room temperature after application of a negative or positive bias to the field plate. The direction of this displacement was opposite to that caused by ion migration or polarization. Further, this phenomenon occurred only after the applied voltage exceeded some critical threshold voltage or field. Various investigators have confirmed this effect with a considerable range of threshold field values. It is generally held that the phenomenon is due to trapping of carriers in the insulator.

Evidence of hole injection and trapping in silicon nitride films prepared by reactive sputtering from an analysis of the C-V hysteresis patterns was presented by Hu et al. (1). Shifts in the C-V characteristic after bias-temperature stress at 300°C support this finding.

The internal field-ionization phenomenon that causes the observed conductivity is also responsible for the charge storage in Si_3N_4 when intense electric fields are employed. This in turn leads to hysteresis in the capacitance-voltage behavior of MNS (metal-nitride-silicon) capacitors as shown in Figure 49. This hysteresis is particularly severe, as shown in Figure 50 according to Gregor (3), above a characteristic threshold field.

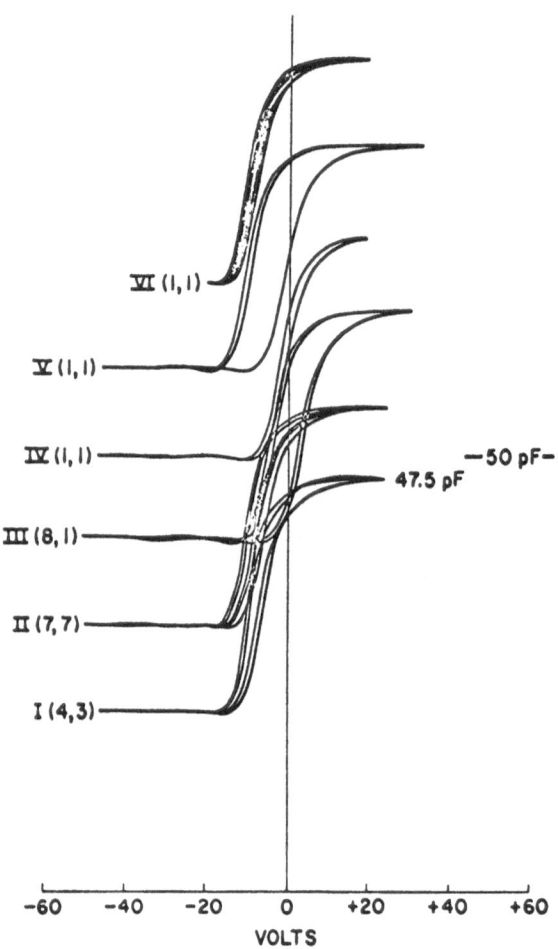

Figure 49. C-V Hysteresis for a MNS Capacitor.

Effect of small field shown in VI; effects of large negative field shown in IV; effects of large positive and negative fields shown in I, II, III, V.

(Gregor (3))

84

It is interesting to note the marked effect of the thin oxide layer in suppressing
the hysteresis as shown in Figure 50. Thus, the use of silicon nitride directly
on a semiconductor surface does not appear to have any practical utility in view
of this hysteresis behavior

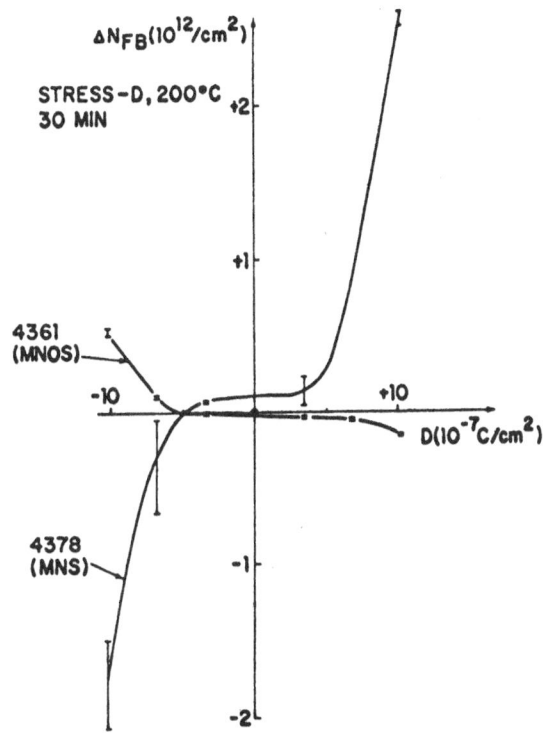

Figure 50. Hysteresis of MNS and
MNOS Capacitors.
Effective DN_{fb} vs.
Electric Deplacement, D.

(Gregor (4))

The usefulness of silicon nitride films for silicon devices is determined, to
a large extent, by the nature of electrical charge distributions in the nitride
film and at the silicon-silicon nitride interface, according to Chu et al. (1).

Chu et al. (1) studied the C-V behavior of a number of $Al-Si_3N_4-Si$ (MNS) capacitors to determine the charge properties in the interface region of $Si-Si_3N_4$
structures. The amorphous silicon nitride films were approximately

85

0.1 micron in thickness on n-type silicon substrates and had 0.033 mm^2 Al electrodes evaporated on the nitride films. Capacitance measurements were made at 100 kHz and 1 MHz. The crystallographic orientation of the substrate surface [111] or [100] and the reactant composition, were found to have no significant effects on the C-V relations. The annealing of the nitride films deposited at 950°C in a nitrogen atmosphere at 1100°C also produced no appreciable changes in the C-V behavior of the MNS capacitors. On the other hand, the addition of minute quantities of oxygen to the reactant gas mixture, changes the electrical properties of the resulting capacitors in a reproducible manner. These authors concluded that the charge instability in MNS structures is a basic property of the Si-Si$_3$N$_4$ structure prepared by the chemical deposition technique and that this property is relatively insensitive to the process parameters investigated by them.

The voltage-current and voltage-capacitance characteristics of MNOS capacitors were investigated by Raburn and coworkers at the University of Alabama, over a wide range of temperatures, to determine the effects of fast and slow interface states on the performance of MNOS devices. Their efforts followed the model proposed by Frohmann-Bentchkowsky and Lenzlinger (1), which assumes that the conduction is due to tunneling from the silicon into the oxide conduction band then drift to the nitride; conduction in the silicon nitride is due to repeated excitations from traps into the nitride conduction band and accumulation of either positive or negative charge at the oxide-nitride interface.

Nitride thicknesses of 500 Å and 2000 Å were used while the oxide thickness (thermally grown), was about 50 Å; both p- and n-type silicon substrates were used. I-V curves were measured for the various capacitor structures at 77, 300 and 423°K. From these measurements, it appears that the conduction mechanism in the nitride is primarily the Frenkel-Poole effect with an ohmic contribution at low fields. At high fields, the shape of the curves for field emission and Frenkel-Poole are frustratingly similar, so that field emission is not completely ruled out. They believed that the nitride conduction mechanisms so completely overshadowed those of the thin oxide layer that nothing concrete can be said about the latter. The C-V hysteresis loop exhibited none of the predicted distortions due to fast surface or interface states, indicating these will not be a problem in memory applications.

RADIATION HARDENING APPLICATIONS

Two approaches to the fabrication of radiation-hard dielectrics, utilizing silicon nitride, have been studied; silicon nitride sandwich structure and silicon oxynitride. Radiation-resistant devices investigated to date have included capacitors (MNS and MNOS), MNOS-FETs, bipolar transistors and operational amplifiers.

Hutchins [1] in a PhD thesis study at North Carolina State University, studied the effects of gamma radiation on MNS capacitors. It was difficult to interpret the data because the radiations were made under short-circuit conditions and the C-V measurements were made some time after terminating the radiation; no stored charge was found. MNS capacitors with 1000 $\overset{\circ}{A}$ of nitride on n-type silicon (2-5 ohm-cm), were irradiated with gamma radiation under dc biases up to +20 V with total doses up to 6 x 10^6 rads (Si) by Perkins [1, 2] at Hughes Aircraft Co. Conductance-voltage measurements were employed to determine the distribution of fast interface states in these MNS capacitors. The samples showed no changes due to irradiation and indicated that the MNS structures appeared to be radiation-hard. The author indicated that the test data was preliminary in nature. Zaininger fabricated MNS capacitors (using SiH_4 and NH_3 reacting gases and nickel for the gates) and studied the effects of high-energy electron bombardment by means of C-V (at 1 MHz) and G-V at (10 kHz) measurements. The bombardment was carried out in air or vacuum. No experimental data were presented in the paper but the statement was made that "electron bombardment of silicon nitride MIS capacitors and transistors, results in effects similar to those for MIS devices with SiO_2 films, but the radiation sensitivity of these structures seems to be much more dependent on fabrication procedures and parameters." The author has proposed a model to describe the radiation effects, noting that the impinging electrons create secondary and tertiary electrons in the insulating films. The trapping of either electrons or holes could then leave a net charge in the insulator. Further, the radiation sensitivity of MIS structures depends on the defect structure of the insulator, and thus on the method of fabrication and annealing and/or ambient treatment.

Perkins et al. (1, 2) used a sandwich structure in which a 125 Å silicon dioxide film is interposed between the silicon and the silicon nitride film in view of the well known charge instability of silicon nitride inherent in MNS structures. (Chu and coworkers had previously shown that by interposing a deposited silicon dioxide layer between the silicon nitride and the silicon, this charge instability can be suppressed to a degree dependent on the thickness of the oxide). Their MNOS capacitors were fabricated on 10 ohm-cm, n-type silicon substrates by first thermally growing a very thin layer of silicon dioxide in dry oxygen at 1000°C, followed by a pyrolytic deposition of silicon nitride to a thickness of 1000 Å. Such capacitors were irradiated in a cobalt-60 source for accumulated doses up to 10^6 rads (Si), with bias voltages of ±5 V. The radiation-induced shifts of the C-V curves along the voltage axis were measured for various thicknesses of the underlying oxide and illustrated in Figure 51.

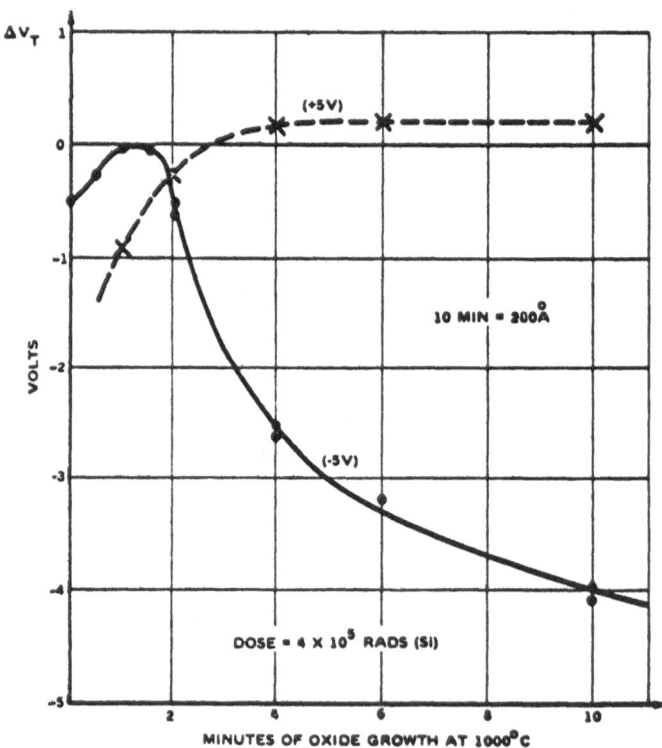

Figure 51. Radiation-Induced Shifts in Threshold Voltage for MNOS Devices vs. Oxide Thickness.

(Perkins et al. (1))

88

The graph indicates an optimum thickness for radiation hardness somewhere in the vicinity of 2 minutes of oxide growth at 1000°C (approx. 125 Å). Additional work was carried on with various collaborators (Perkins et al. (2)) by growing an oxide thermally in dry oxygen at 1000°C to various thicknesses before applying the silicon nitride layer. These MNOS structures were evaluated as capacitors and irradiated in a cobalt-60 source with ±5 V bias applied remotely. After periodic intervals, the capacitors were withdrawn from the source and the capacitance vs. voltage (C-V) measured (frequency of carrier voltage was 10 kHz) as plotted in Figure 52.

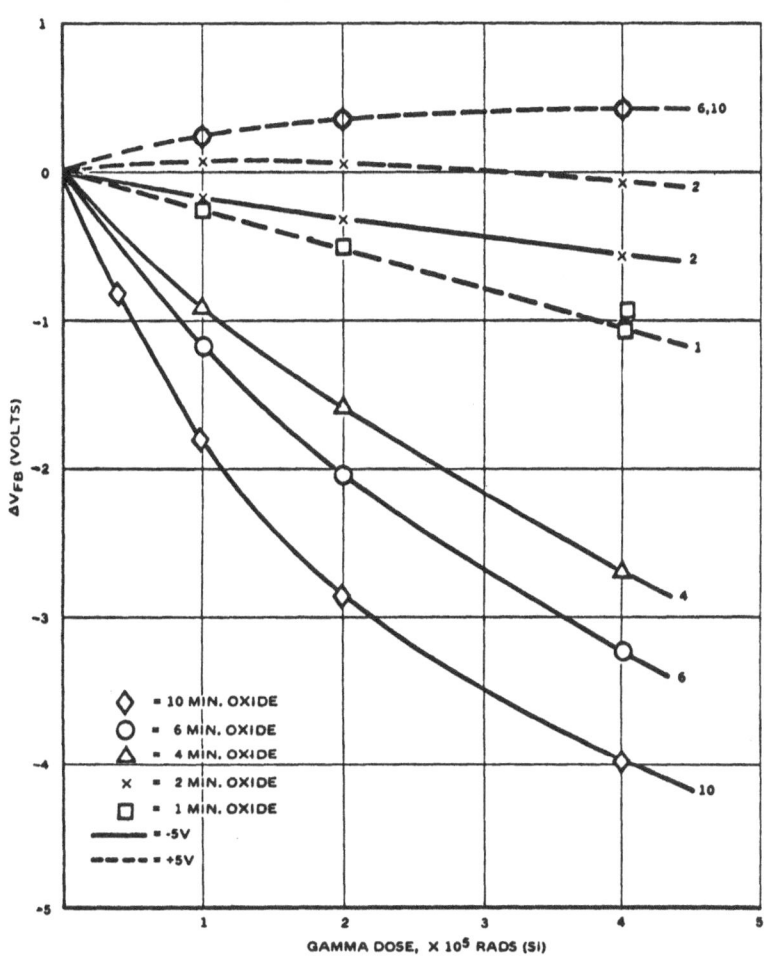

Figure 52. C-V Curve Shifts as Function of Dose for MNOS Capacitors at ±5 V.

(Perkins et al. (2))

These results show that the silicon nitride produces a decrease in radiation sensitivity under positive biases and an increase (quite drastic), under negative biases. It was discovered moreover, that by making the oxide thin, the sensitivity under negative biases could also be decreased. The curves show that for oxide thicknesses of the order of 100 Å, the effects are small at both polarities. The authors concluded that MNOS structures were radiation hard or resistant.

Cricchi and Ferber reported at the Oct. 1970 International Electron Devices meeting on the development of low-threshold-voltage radiation-resistant CMNOS arrays utilizing a multilayer nitride-oxide gate insulator with thicknesses of 750 Å and 50 Å, respectively. Deposition of the nitride at 875°C provided an initial interface charge density of $\pm 5 \times 10^{10}/cm^2$. Long-term stability tests at 150°C indicated that negligible shifts in threshold voltage occur for the low electric field (from 1 to 2 x 10^6 V/cm) experienced during normal operating conditions. CMNOS memory arrays were also tested to determine their radiation resistance. The inherent photocurrent compensation in symmetrical complementary memory arrays, the high noise immunity nitride-oxide memory cell, and the low field conditions for the multilayer nitride-oxide gate insulator, tend to improve the radiation resistance, according to these investigators.

In 1966, NASA investigators (Gordon and Wannemacher) studied the effects of space radiation (electron fluences) on experimental p-channel enhancement MNS-FET devices (made by Sperry Rand Research Center) and compared them with MOS-FETs. The trends were favorable for the MNS-FETs, however much more physical and engineering data collection and analysis was required to come to definite conclusions regarding their radiation resistance behavior. Subsequent studies were carried out by Newman and Wegener on p-channel enhancement types with a layer of oxide at the interface. The basic MNS device has a 1850 Å layer of silicon nitride (formed by the thermal decomposition of SiH_4/NH_3) on a chemically cleaned, 10 ohm-cm, n-type silicon wafer as shown in Figure 53. Newman and Wegener summarized the effect of radiation on the gate threshold property of MNS devices manufactured by several semiconductor companies as shown in Figures 54 through 58 and in Table 6.

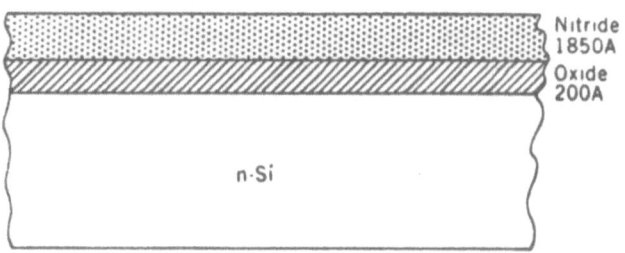

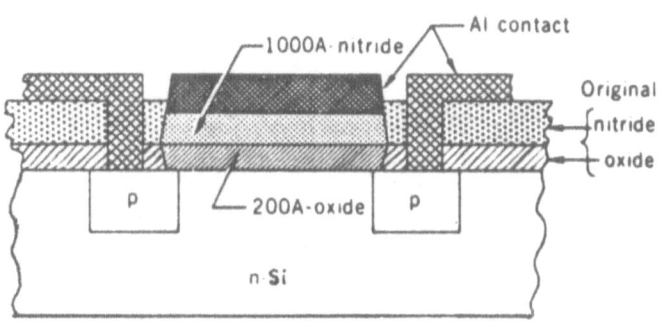

Figure 53. Metal-Nitride-Semiconductor Field
Effect Transistor Structure.
(Newman and Wegener)

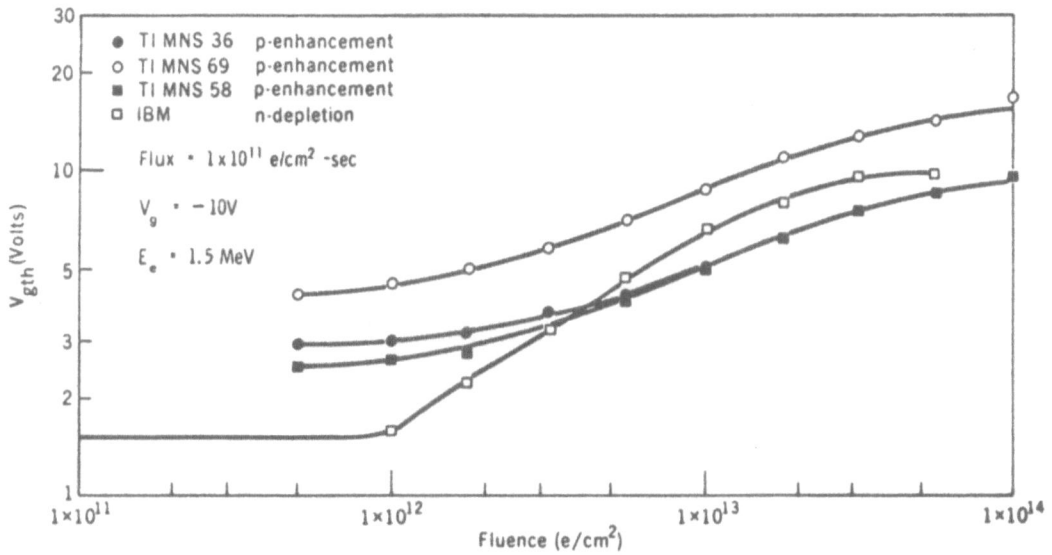

Figure 54. Gate Threshold Voltage vs. Fluence for TI and
IBM MNS Devices.   (Newman and Wegener)

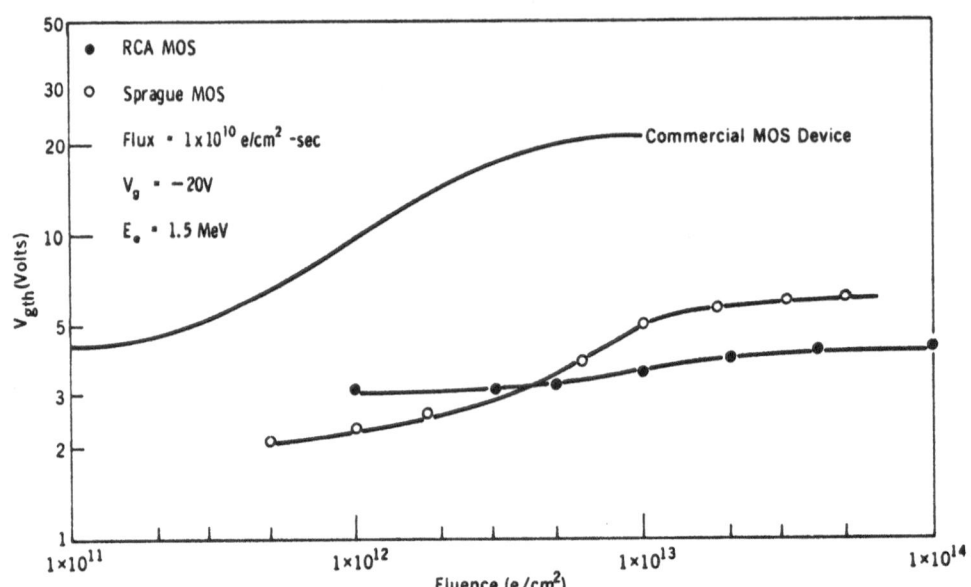

Figure 55. Gate Threshold Voltage vs. Fluence for Sprague
and RCA MOSFET Devices. (Newman and Wegener)

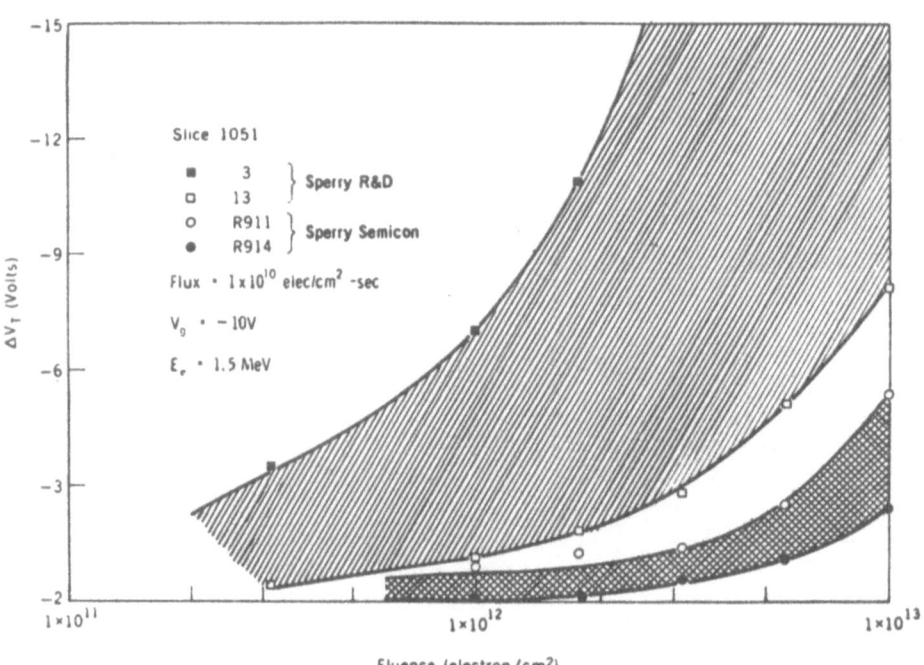

Figure 56. Change in Gate Threshold Voltage
vs. Fluence for MNS Devices Packaged under Different Conditions.
(Newman and Wegener)

92

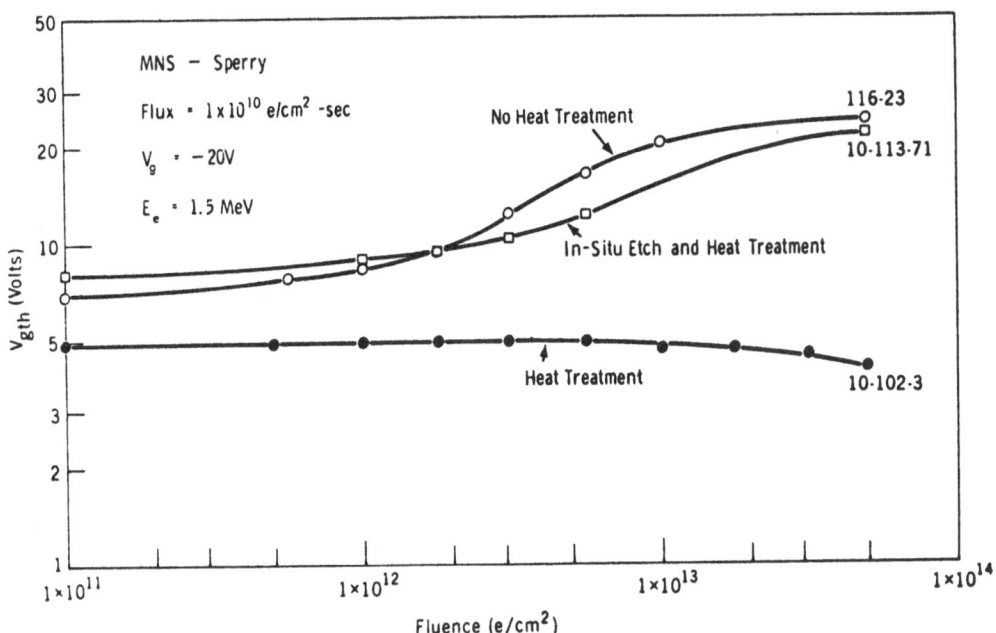

Figure 57. Gate Threshold Voltage vs. Fluence for MNS
Heat Treatment and In-Situ Etching.
(Newman and Wegener)

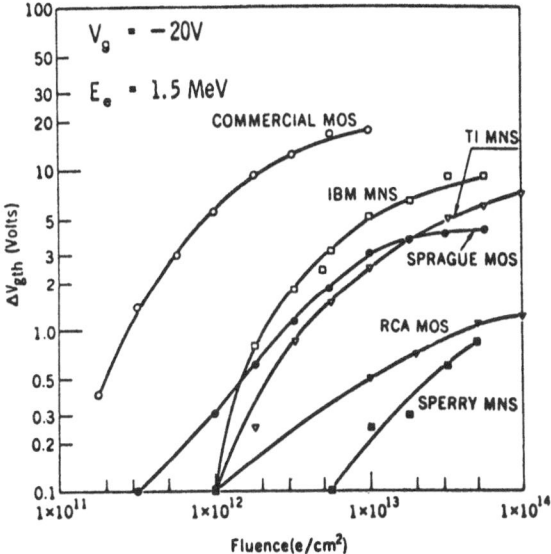

Figure 58. Gate Threshold vs. Fluence
for Commercial MOS and
Representative Radiation
Tolerant Devices.
(Newman and Wegener)

Table 6. Radiation Tolerant MISFET Devices. (Newman and Wegener)

Fluence	1×10^{12}	5×10^{12}	1×10^{13}	V_{to}	Thermal Stab.
Sperry 10-102-3	0 v	.1 v	.25 v	-5.0 v	.25 v
Sperry 10-102-88	.1	.6	.8	-4.7	.5 v
RCA	-.1	-.5	-1.2	-3.0	.3 v
Ti 58	-.1	-.7	-2.5	-2.5	not meas.
IBM 56	-.1	-2.4	-5.2	-1.5	1 v
Sperry 10-51 R911	-.1	-1.2	-2.4	-7.1	.5 v
Sperry 10-51 R912	-.1	-2.2	-4.4	-7.2	.3 v
TI 52	-.3	-1.0	-1.8	-1.9	not meas.
Sprague TXF200-3	-.3	-1.5	-2.8	-2.0	.5 v
TI 69	-.3	-1.3	-2.3	-2.7	.5 v
IBM 57	-.4	-.4	-1.4	-1.2	.5 v
Sprague TXF200-4	-.4	-2.0	-3.3	-2.0	.5 v
Sperry 10-102-5	-.6	-1.8	-2.5	-8.8	.2 v

Again, no general conclusions were possible because of not enough sub-
stantiating data.

In 1967, Totah at Hughes Aircraft Co. reported the results of his NASA contract study involving the effect of radiation on the stability of silicon nitride and metal-nitride-semiconductor FET devices. The latter, exposed to ionizing radiation, exhibited only a slight shift in the threshold voltage (V_T). Such devices should be able to operate satisfactorily when exposed to radiation doses of at least 10^6 rads. MOS devices, evaluated for comparison purposes, showed significant changes in V_T above 10^4 rads. Additionally, exposure to the electron source resulted in a maximum 25% etch rate enhancement.

Perkins and coworkers (1, 2) at Hughes Aircraft Co. briefly investigated the MNOSFET structure for radiation hardness. The electrical stability was measured for the transistors by applying a dc stress voltage to the gate for two minutes and then removing it in order to measure the threshold voltage. Figure 59 shows a typical result for positive stress on the gate with subsequent recovery using a negative stress. This behavior agreed with results by Chu et al. and Deal et al. according to these authors.

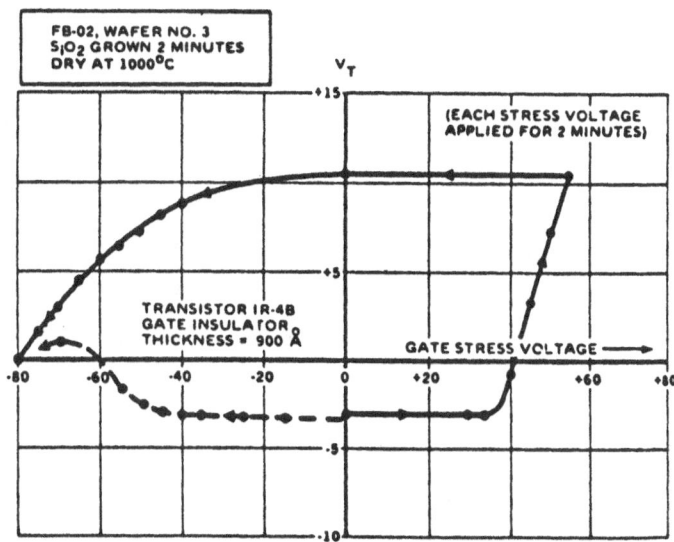

Figure 59. Threshold Voltages of MNOS Transistors after Successive Applications of Increasing Stress Voltages at the Gate.

(Perkins et al. (1))

The figure indicates the region of electrical stability extends from gate voltages of -40 V to +35 V. For thicker oxides, the width of this stable region broadens somewhat.

As early as 1966, A.G. Stanley reported on the effects of 1 MeV electron irradiation on metal-nitride-semiconductor insulated gate field-effect transistors (IGFETs) over a wide range of biasing conditions. In such devices, the radiation induced shift in gate turn-on voltage was greatly reduced and source-drain leakage currents eliminated. He concluded that such MNS devices can operate satisfactorily under ionizing radiation after radiation hardening. In another paper, he evaluated the effect of silicon nitride vs silicon dioxide and SiO_2-Si_3N_4 composite passivation on the performance of bipolar transistors which were subjected to 20 keV and by 1.5 MeV electron radiation. The results showed that silicon nitride and silicon dioxide passivated transistors possess identical properties, after irradiation by low energy electrons to saturation; irradiation by 1.5 MeV electrons caused channel formation in all three groups. A number of graphs and tables compare the radiation effects on silicon nitride alone, silicon nitride with silicon dioxide, and silicon dioxide alone.

A monolithic radiation-hardened operational amplifier has been designed and built by Stafford and Oberlin at Fairchild Semiconductors, utilizing silicon nitride along with dielectric isolation processing. The device was tested in a variety of radiation environments, including fast neutrons, flash x-rays and cobalt-60 sources. Circuits were exposed to fast neutron doses from 3×10^{12} n/cm^2 to 1×10^{14} n/cm^2. The only electrical parameters showing significant variations with radiation were open-loop gain and input bias current. The voltage gain was still within allowable limits after a dose of 3×10^{13} n/cm^2, and the amplifier still had useful gain at 10^{14} n/cm^2. The input bias current and input resistance decreases to about 70 k-ohms after exposure to 10^{14} n/cm^2.

Some interest has been shown in silicon oxynitride films in view of its showing greater radiation resistance than conventional silicon dioxide films. According to Schmidt and coworkers at Bell Telephone Laboratories, a silicon oxynitride (SiON) in MIS structures with composition near Si_2ON_2 is nearly insensitive to space-charge build-up for radiation doses up to approximately 10 Mrads. Bias during irradiation does not lead to instability unless certain threshold field values are exceeded. These values are approximately 2.8×10^5 V/cm for negative bias and approximately 1.5×10^6 V/cm for positive bias. Above the threshold for negative bias there is negative charge injection by the metal contact. The shift of the C-V curves, ΔV_{fb}, is linearly related to the applied bias (V_A) by the equation;

$$\Delta V_{fb} = V_{TH} - V_A$$

where V_{TH} is the threshold voltage. They concluded that SiON films appear to be suitable for use in planar devices in radiation environments. However, in order to achieve low V_{fb} and N_{fs} values, the SiON films must be separated from the silicon surface by a thin SiO_2 layer, or the silicon surface must be specially prepared.

In a second communication, Schmidt and Ashner report on additional experiments carried out with silicon oxynitride films in capacitor structures, passivated bipolar transistors, and p-channel IGFET structures. B-T (Bias-Temperature) tests, under negative bias at 1.6×10^6 V/cm and at temperatures up to 330°C showed a flatband voltage (V_{fb}) drift of 2 to 3 V to more positive values during the first hour, which did not change thereafter, very little drift occurred under positive bias and the same test conditions. Moreover, the B-T stability, observed on Si/SiON/Al structures, is often lost after gold wire thermo-compression bonding. The authors came to the following conclusions with respect to their silicon oxynitride results:

1. SiON passivation of silicon surfaces, followed by an anneal in hydrogen at 900°C for 1/2 hr. results in interface state densities in the 10^{11} $eV^{-1}cm^{-2}$ range over most of the forbidden gap. These densities remain practically unchanged under exposure to ionizing radiation in the several Mrad range.

2. The B-T stability of SiON films under positive bias is good; under negative bias it is marginal (2-3 V drifts at 1.6×10^6 V/cm and 330°C), but probably acceptable for devices whose primary function is to be radiation-hard at or near room temperature.

3. The permissible dose of ionizing radiation for SiON passivated bipolar transistors (type 16F NPN) is about 30 times greater than for SiO_2 passivated bipolars. The increase in radiation-hardness can probably be achieved without compromising any other desirable device characteristics. Power-aging at 850 mW (junction temperature 300°C) showed that SiON passivated bipolars were as reliable as good quality SiO_2 passivated devices; the median life was 700 hours. Radiation-hardness was retained after power-aging.

4. Fabrication of radiation-hard IGFETs using SiON as the gate insulator is feasible but subject to the condition that a negative bias applied to the gate during irradiation should not exceed the threshold field for NBI (4×10^5 V/cm in the most favorable case of SiON deposited on fresh epitaxial silicon). These devices possess the same hole drift mobility (200 cm^2/V sec) as SiO_2 passivated devices of the same geometry.

5. Problems persist with the initial position of V_T and with the threshold field to drifts under irradiation with negative bias which at present prevent the use of SiON passivated IGFETs in digital circuits. These devices could be used in linear circuits.

6. Overcoming these problems depends in part on the solution or circumvention of an impurity diffusion problem arising during the fabrication of SiON passivated p-channel IGFETs with small negative V_T values. Impurity diffusion apparently lowers the threshold field to NBI below the value given in paragraph 4 above. In the worst case, the threshold can disappear. Chances appear good that the impurity problem could be solved with some additional work.

7. Diffusion of gold at 900°C, 1/2 hr., into Si/SiON structures resulted in the introduction of a negative charge into the silicon surface, and a radiation instability under positive bias or without bias. Diffusion of copper does not cause these effects. Thus a lifetime killer other than gold may be preferable for suppression of the p-n junction photoeffect in radiation-hardened IGFET circuits.

Silicon nitride has been explored in MNST to aid in the quantity and stability of the charge which forms at the interface and in the insulator or gate dielectric. Tombs et al. have described a planar silicon insulated-gate, field-effect transistor that utilizes silicon nitride as a diffusion mask, a passivating layer and as the gate insulator. They claim that use of this material in place of the conventional, thermally-grown silicon oxide has indicated improvements in the stability of the device, the dielectric strength of the gate, and the control of the surface state density. Amorphous silicon nitride films were utilized for the above device with a resistivity of 5×10^{14} ohm-cm at room temperature.

The use of sputtered silicon nitride as part of the gate insulation of cadmium sulfide, thin film transistors (TFT) has been reported to provide greater device stability than SiO_2 or SiO, according to Neugebauer and coworkers.

Silicon nitride appears to be attractive for use in MIS devices, according to Lamb, for several reasons: (1) it is a better barrier against diffusants and ionic drift than the oxide, (2) it has a higher dielectric strength than the oxide, (3) it has a higher dielectric constant than the oxide, which would improve the beta factor of an MIS transistor. Nitride film thicknesses ranging from 400 to 4000 Å were prepared by reacting silane (SiH_4) and ammonia (NH_3) at 850°C, using nitrogen as a carrier gas, on float-zone (111), 1 to 10 ohm-cm, p-type silicon. The resulting Si_3N_4 films had a dielectric constant of 6.7 ±0.5 and an etch rate of 100 Å/min. in 40% HF at room temperature. Analysis of the electrical properties (e.g. C-V measurements), shows that a marked hysteresis effect occurs. This hysteresis appears as a shift in the accumulation/depletion region of the capacitance curves along the voltage axis in the direction of the applied electric field, and also as a change in shape of the capacitance curve; both changes being a function of the rate of change of the applied field. He concludes that the high surface charge and the hysteresis effects prevent the general use of silicon nitride in MIS transistors. A private communication by Brotherton notes that the magnitude of the hysteresis is a function of the silicon orientation and decreases in the order (111)>(110)>(100).

Minority carrier injection using metal-insulator-semiconductor (MIS) structures driven by ac excitation has been achieved for both n- and p-type GaAs/Si_3N_4 devices and luminescence is observed. The operation depends on the ability of the GaAs to produce minority carriers in the space-charge region during the positive half-cycle of the voltage. The advantage of such a device is that electroluminescence can be obtained where p-n junctions are either difficult or impossible to fabricate. Gallium arsenide samples were coated with a 1000 Å film of Si_3N_4. Even infrared radiation was observed using a special setup. Uniform light generation in both n- and p-type GaAs over the area of the fabricated MIS capacitor could be observed when peak voltages of the applied ac signal exceeded the avalanche point.

Gallium arsenide insulated-gate, field-effect transistors have been developed by Becke and White using silicon nitride as the gate dielectric. Compared with metal-silicon dioxide-gallium arsenide structures, the metal-silicon nitride-gallium arsenide devices proved superior due to a lower interface state density ($<10^{12}/cm^2$ eV), the higher dielectric constant of the nitride films and the greatly improved diffusion masking against zinc and tin dopants. Capacitance-voltage curves for SiO_2 and Si_3N_4 on GaAs show a large relative change of capacitance with voltage for silicon nitride and a significant improvement over silicon dioxide as shown in Figure 60. The large change in capacitance for silicon nitride on GaAs indicate the field is penetrating the semiconductor. The silicon dioxide curve is relatively flat, showing the field lines terminate at the interface of the insulator and substrate, and do not affect charges that are deeper.

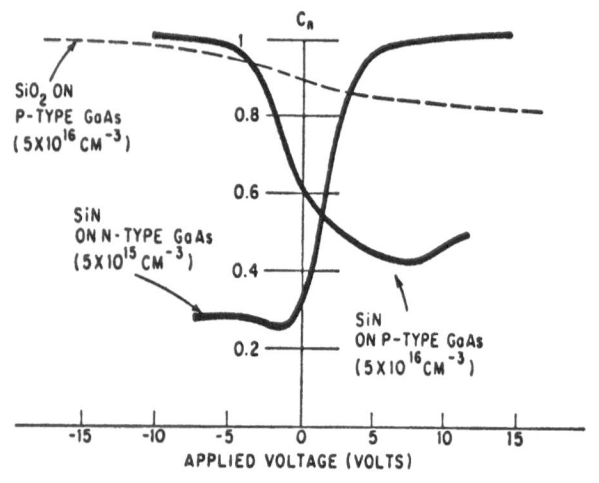

Figure 60. Change in Capacitance for Silicon Nitride on Gallium Arsenide.

(Becke and White)

100

Foster and Swartz have investigated the gallium arsenide MNSFET device utilizing silicon nitride as the gate insulator as well as the electrical characteristics of the Si_3N_4-GaAs interface. The silicon nitride films were pyrolytically deposited from silane (SiH_4) and ammonia (NH_3) at 700-725°C. After deposition of the Si_3N_4, aluminum dots were evaporated onto the chip through a metal mask. The dots were nominally 10 to 20 mils in diameter and 10,000 Å thick. The silane/ammonia gas flows were varied so that different Si_3N_4 film thicknesses (800 to 1200 Å) were obtained. Both n- and p-type GaAs substrates were evaluated. The films were evaluated by C-V curves and considerable differences found in the various samples, especially with respect to the NH_3/SiH_4 ratios and the p-type vs n-type samples. The C-V curves for n-type substrates are shifted to the right, giving positive flat-band voltages in contrast to negative flat-band voltages for p-type samples. This strongly suggests that the surface states are negative when below the Fermi level, thus accounting for the difference between the curves on p- and n-type substrates. The positive charge causing the shift of C-V curves on p-type substrates must, then, be accounted for by traps that are positive when empty. Their studies have revealed that the predominant characteristic of the Si_3N_4-GaAs interface is the presence of hysteresis due to the filling and emptying of traps in the nitride near the interface. The spatial distribution of these traps from the interface is strongly dependent on the NH_3/SiH_4 ratio. The amount of this hysteresis was minimized to less than 2 V for fields below 1.7×10^6 V/cm for the best p-type sample. Best results were obtained at a deposition temperature of 700°C and a NH_3/SiH_4 ratio of 62.5/1. Surface state density for the best p-type sample was in the 10^{12} to 10^{13} range. Some further reduction in the amount of hysteresis is needed, however, before stable gates can be fabricated from GaAs-Si_3N_4 FETs according to these investigators.

Sperry Rand Corp. (Soref) has explored the possibility of fabricating high-gain, low noise photodetectors utilizing silicon nitride layers (2000 Å thick) deposited on silicon substrates with semi-transparent aluminum grid electrodes to detect 10.6 micron laser energy. A cross-section view of the device is illustrated in Figure 61.

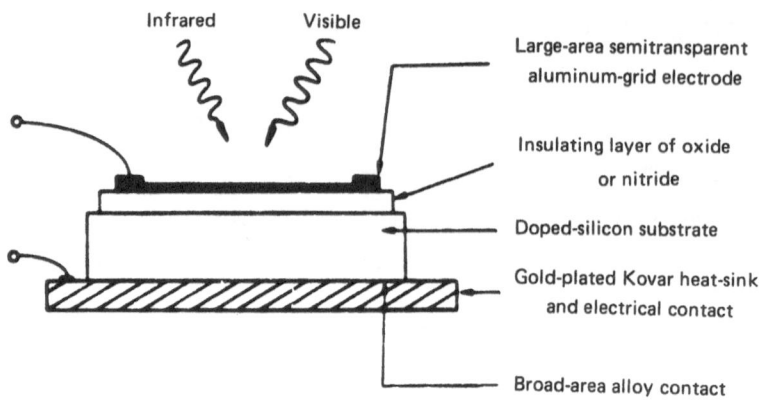

Figure 61. Cross-Section View of the MIS
Infrared Image Sensor.

(Soref)

Present evidence indicates that the mechanism of detection is impurity photo-conduction in both the n-type silicon substrate and the diffused p-type regions.

Tombs et al. in 1966 demonstrated the feasibility of using silicon nitride as the dielectric in insulated-gate, field-effect transistors. Sarace and co-workers at Bell Telephone Laboratories reported on MNOS silicon insulated-gate, field-effect transistors fabricated by processes involving relatively non-critical photoresist and self-limiting etching steps. The important features of their method included the formation of the gate insulator under extremely clean conditions, incorporation of an alkali ion barrier of silicon nitride to achieve stable device characteristics and automatic alignment of the gate electrode with respect to source and drain. The gate insulator comprised 600 $\overset{\circ}{A}$ of thermally grown Si_3N_4 which are formed at the beginning of fabrication. A thick 8000 $\overset{\circ}{A}$ layer of SiO_2 is pyrolytically deposited over the nitride to minimize contact capacitances in the finished structure. The nitride layer serves the dual function of providing a barrier to mobile ions during fabrication to achieve control over geometry. The complete structure is illustrated in Figure 62.

In 1969, General Instrument Corp. announced a low voltage nitride MTNS device having a metal-thick oxide-nitride-silicon structure as shown in Figure 63.

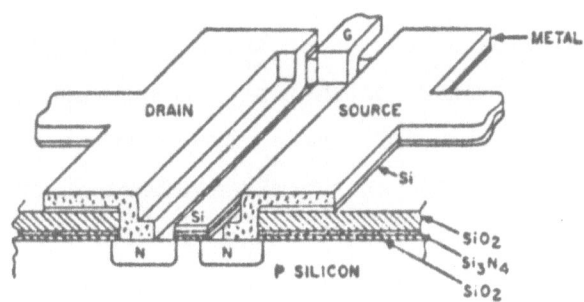

Figure 62.   Final Geometry after
              Metallization and
              Diffusion.

                    (Sarace et al.)

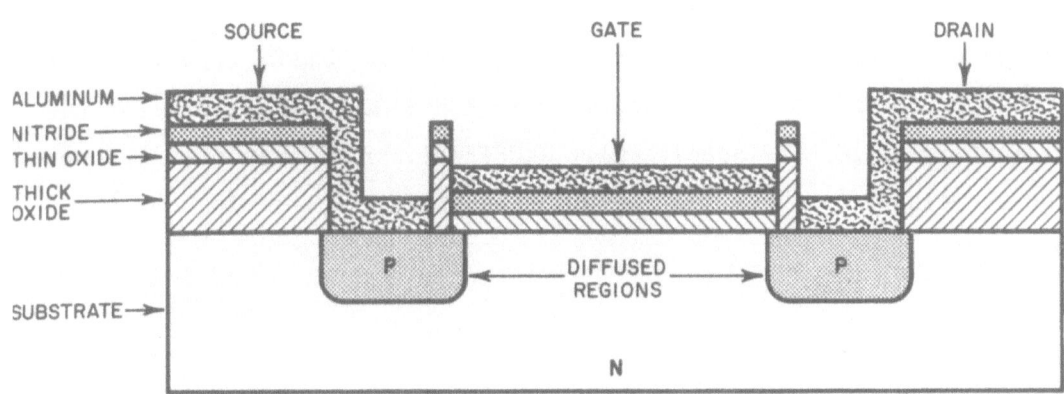

Figure 63.   Low Voltage MTNS Device
          (General Instrument Corp.)

According to the company, the addition of silicon nitride to the gate structure
of the device has the following beneficial effects on the basic reliability of
the MTNS structure: The first and most important is that it makes the device
virtually impervious to sodium ion migration, thereby greatly enhancing the
long term stability of threshold voltages and leakage currents. Because of
this, devices are typically designed and rated for operation at 125°C. Secondly

there is a reduction in the magnitude of the supply voltages required; only a +5 and a -12 V supply is required. Another consideration is that the low voltage circuits dissipate less power than the standard voltage circuits; and therefore have less internal heating which keeps junction temperatures at a minimum. Thirdly, the need for a great many interface circuits is eliminated. These latter circuits and their associated interconnections often contribute a greater proportion of unreliability than the LSI circuits. A reliability report is available summarizing the testing and resultant failures obtained in a number of discrete devices including a MTOS Zener protected single device, manufactured using the nitride process. Additional details and criteria for deciding whether or not to use this MTNS device were presented in a paper given at the 1969 NEREM Conference by Zimbelmann and Penner of the same company.

In August 1970, General Instrument Corp. introduced a pair of MTNS (metal-thick oxide-nitride-silicon) chips that function as a computer terminal receiver and a transmitter. They are available in 24-lead dual in line packages.

Carlson and coworkers at Texas Instruments investigated MNOS capacitor structures in order to understand the use of combined SiO_2-Si_3N_4 layers for the gate dielectric of IGFETs (specifically MNOSFETs). Figure 64 defines one of the process methods used for the formation of MNOSFET transistors by this company.

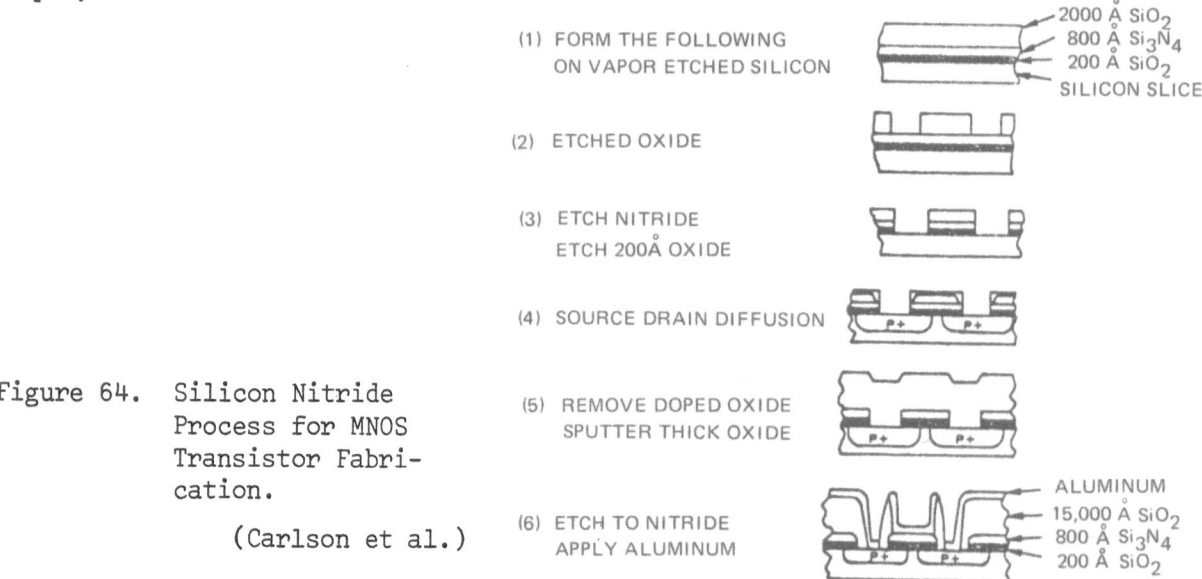

Figure 64. Silicon Nitride Process for MNOS Transistor Fabrication.

(Carlson et al.)

104

The original report should be consulted for details on the etchants employed, deposition temperatures and conditions and diffusion conditions. Another process used by them for fabricating MNOS field-effect transistors is illustrated in Figure 65. This latter process involves deposition of the nitride film near the end, rather than the beginning of fabrication. The first process results in an inherently more stable structure because of the sealing of the surface early in the process by the silicon nitride deposition. The second process is more sensitive to contamination in processing, but the finished device is protected by a smooth coat of silicon nitride over everything but the contacts.

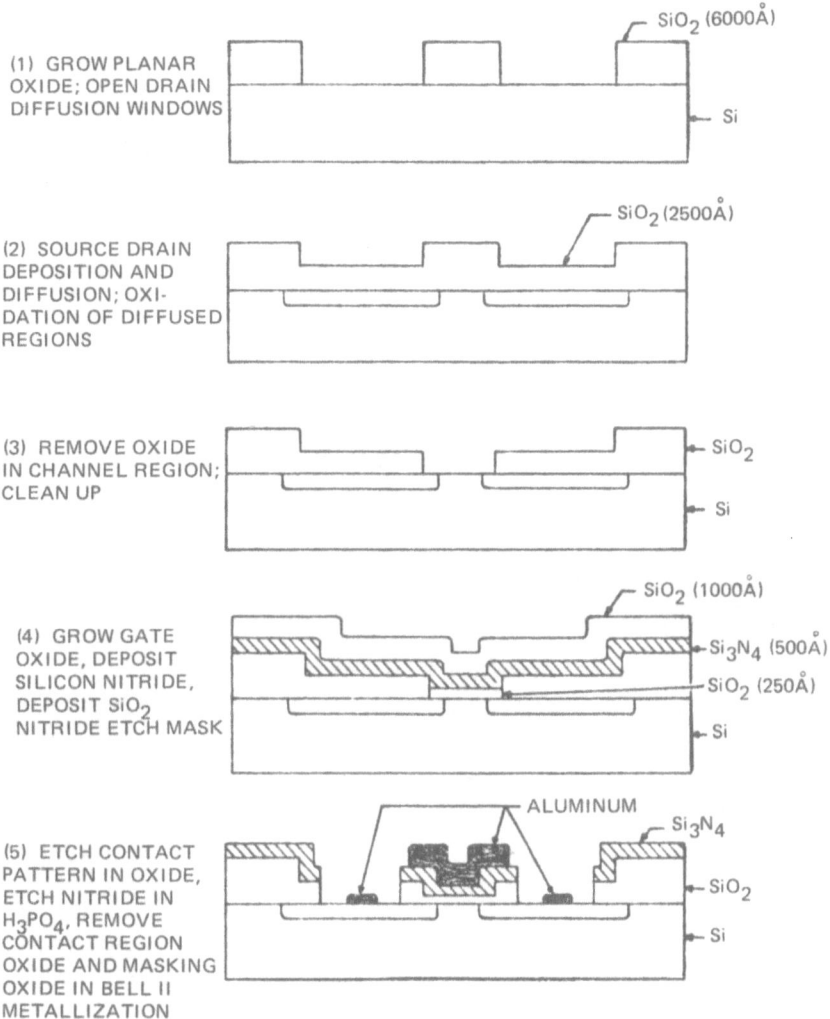

Figure 65. Second Process for Fabricating MNOS Field-Effect Transistors.

(Carlson et al.)

Hillery also investigated MNOS structures under various combinations of temperature-bias stressing conditions and came to the following conclusions:

1. The heating and cooling cycles involved during the nitride deposition process can affect the characteristics of the underlying oxide.

2. The use of an inert gas for these cycles minimizes the effect on the Si/SiO_2 interface.

3. A second, and more pronounced, source of excess charge results from the properties of the nitride film itself. It is believed that this effect is caused by the presence of a silicon-rich layer at the SiO_2/Si_3N_4 interface.

4. This effect also can be minimized by the selection of a gas cycling sequence, compatible with the deposition system used, which will not allow the silane to dissociate over the wafers in advance of the dissociation of the ammonia.

The development of a dual 25-bit, 4-phase shift register using MNOS structures was reported in 1967 by Keshavan et al. (1) of Westinghouse Electric Corp. Threshold voltages of 1.5 to 2 V have been achieved using pyrolytic oxide-nitride and thermal oxide-nitride. It was found that MNOS shift-registers with pyrolytic oxide is more stable than with thermal oxide. Switching speeds of the shift-register in the low MHz was achieved. Radiation resistance of MNOS and MOS shift-registers were also compared by these investigators.

Low-threshold-voltage, radiation-resistant, complementary metal-nitride-oxide-semiconductor (CMNOS) arrays have been constructed by Cricchi and Ferber, utilizing a multilayer nitride-oxide gate insulator. Both the n- and p-type channel CMNOS devices have threshold voltages that are typically 0.5 to 0.9 V. This low threshold permits operation of the arrays with a 5 V power supply providing very low power dissipation and low insulator electric fields. Their CMNOS devices were fabricated on 5 to 10 ohm-cm, (100), n-type silicon having a p-type well surface concentration of nominally 2×10^{16} cm^{-3}. The nitride-oxide insulator thicknesses were typically 750 Å and 50 Å respectively, giving an effective insulator thickness equivalent to a single 500 Å oxide layer. Deposition of the nitride at 875°C provided an initial interface charge density of $\pm 5 \times 10^{10}/cm^2$. According to these workers, all of these factors combine to provide high-quality, nitride-passivated, low-threshold voltage CMNOS devices. Long-term stability tests at 150°C indicate that negligible shift in threshold voltage occurs for the low electric field (from 1 to 2 x 10^6 V/cm) experienced during normal operating conditions

CMNOS memory arrays were tested to determine their radiation resistance. The inherent photocurrent compensation in symmetrical complementary memory arrays, the high noise immunity of the CMNOS memory cell and the low field conditions for the multilayer nitride-oxide gate insulator all tend to improve the radiation resistance. The CMNOS memory arrays have been exposed to flash x-ray fluences exceeding 10^9 rads/sec with logic error threshold typically between 5 x 10^8 and 10^9 rads/sec.

Affleck et al. at General Electric used a MONOS structure which comprised 1100 Å of thermally grown SiO_2, 1000 Å of pyrolytically deposited Si_3N_4 and 4000 Å of pyrolytically deposited SiO_2 with aluminum contact pads deposited by two techniques, (tungsten filament evaporation and electron beam systems), to study the effects of charge distribution in oxide-nitride layers. The resulting MONOS structure was studied by measuring the capacitance-voltage characteristics both before and after temperature-bias stressing. The tungsten evaporation system indicated contamination in the top layer by showing changes in the values of Q_{ss} at the silicon-silicon dioxide interface. The electron beam evaporation system produced a contact free of ionic contamination although radiation damage is generated during the deposition process. This radiation damage is readily annealed from MOS structures but complete removal has been shown to be much more difficult in the case of MONOS structures.

Multilevel wiring technology, using aluminum and silicon nitride, has been developed to an advanced stage by Standard Telecommunication Laboratories Ltd. (Harlow, Essex, England). The company chose silicon nitride because it is one of the best insulators. To improve adhesion to the aluminum layers, the nitride is placed between two thin films of silicon dioxide. This oxide-nitride-oxide sandwich is deposited in one sequential operation by a glow-discharge-activated vapor deposition process that gives uniform thickness over the steps of the very non-uniform surfaces of IC slices. Both yield and reliability have been explored by first defining multilayer components that can be investigated on especially designed test slices. Three types of components can be identified: conductor tracks, feed-through connections and conductor crossovers. Two types of test slices have been used to evaluate the process. One contains 10,000 crossovers each 25 microns square; the other, 300 feed-through connections of three different sizes; 25, 50 and 75 microns square. Results indicate that a high yield is obtained more readily on crossovers than on feed-throughs. The yield on crossovers is largely determined by the number of pinholes in the insulating films.

Feed-through yield is influenced by:
a. condition of the interface,
b. thickness of the conductor film in relation to that of the insulating film,
c. angle of incidence of the aluminum vapor during deposition,
d. photolithographic process conditions,
e. etch behavior of the insulating film when etching holes.

Compatibility of the $Al-Si_3N_4$ system with SiO_2 interfaces has been investigated at high temperatures over extended periods of time. Results have shown that the system is extremely stable; temperatures up to 600°C have not revealed any reaction of the aluminum with the silicon nitride. Temperature cycling carried out between -50 and +100°C has not revealed any measurable change in electrical or physical properties. No deterioration in crossovers or feed-throughs has been observed after 5000-hour storage at 200°C under bias conditions.

REFERENCES

AFFLECK, J.H. et al. Effects of Charge Distribution in Oxide-Nitride Layers. Abstract no. 111, THE ELECTROCHEMICAL SOCIETY, Extended Abstracts of Spring Meeting, Los Angeles, May 10-15, 1970. p. 288-289.

APPELS, J.A. et al. Local Oxidation of Silicon and Its Application in Semiconductor Technology. PHILIPS RESEARCH REPORTS, v. 25, no. 2, Apr. 1970. p. 118-132.

BARNES, C.R. and C.R. GEESNER. Silicon Nitride Thin Film Dielectric. ELECTROCHEMICAL SOC., J., v. 107, no. 2, Feb. 1960. p. 98-100.

BARNEY, W. Nitride Passivation Links TTL, MOS. ELECTRONICS, v. 42, no. 7, Mar. 31, 1969. p. 117-120.

BARRACLOUGH, M. et al. MIS Memory Transistors. ROYAL RADAR ESTABLISHMENT NEWSLETTER AND RES. REV., no. 9, 1970. Malvern, England. N71-10221. 3 p.

BECKE, H.W. and J.P. WHITE. Gallium Arsenide Insulated Gate Field Effect Transistors. INSTIT. ELECTRICAL ENGINEERS CONF., SERIES no. 3. Gallium Arsenide Symposium, Reading, England. 1966. Paper no. 3, p. 219-227. [1]*

BECKE, H.W. and J.P. WHITE. Gallium Arsenide FETs Outperform Conventional Silicon MOS Devices. ELECTRONICS, v. 40, no. 12, June 12, 1967. p. 82-90 [2]

BROWN, D.M. et al. A New Masking Technique for Semiconductor Processing. ELECTROCHEMICAL SOC., J., v. 114, no. 7, July 1967. p. 730-733. [1]

BROWN, D.M. et al. Refractory Metal Silicon Device Technology. SOLID STATE ELECTRONICS, v. 11, no. 12, Dec. 1968. p. 1105-1112. [2]

BURGESS, T. et al. Thermal Diffusion of Sodium in Silicon Nitride Shielded SiO Films. ELECTROCHEMICAL SOC., J., v. 116, no. 7, July 1969. p. 1005-1008.

CARLSON, H.G. et al. Development of Silicon Nitride and Cermet Resistors, for Use in a Binary Counter, Metal Insulator Field Effect Transistor Circuit. Texas Instruments Report no. 03-68-42, Mar. 1968. 147 p. N69 10283.

CHAIKIN, S.W. Reliable Dielectric Films for Microcircuits. Stanford Research Institute, Final Report, May 1965. AD 617 958.

CHU, T.L. et al. Preparation and C-V Characteristics of $Si-Si_3N_4$ and $Si-SiO_2-Si_3N_4$ Structures. SOLID STATE ELECTRONICS, v. 10, no. 9, Sept. 1967. p. 897-905. [1]

CHU, T.L. et al. The Preparation and Properties of Amorphous Silicon Nitride Films. ELECTROCHEMICAL SOC., J., v. 114, no. 7, July 1967. p. 717-722. [2]

CHU, T.L. et al. Films of Silicon Nitride-Silicon Dioxide Mixtures. ELECTROCHEMICAL SOC., J., v. 115, no. 3, Mar. 1968. p. 318-322. [3]

* Number in parentheses corresponds to number shown in text.

CHU, T.L. et al. Interface Characteristics of Silicon-Si_3N_4 Structures. THE ELECTROCHEMICAL SOCIETY MEETING, Philadelphia, Penna., Oct. 1966. [4]

CRICCHI, J.R. and R.R. FERBER. Low Threshold Voltage Radiation-Resistant CMNOS Arrays. INTERNATIONAL ELECTRON DEVICES MEETING, Washington, D.C. Oct. 1970.

DALTON, J.V. and J. DROBEK. Structure and Sodium Migration in Silicon Nitride Films. ELECTROCHEMICAL SOC., J., v. 115, no. 8, Aug. 1968. p. 865-868.

DEAL, B.E. et al. Characteristics of Fast Surface States Associated with SiO_2-Si and Si_3N_4-SiO_2-Si Structures. ELECTROCHEMICAL SOC., J., v. 116, no. 7, July 1969. p. 997-1005. [1]

DEAL, B.E. et al. Electrical Properties of Vapor-Deposited Silicon Nitride and Silicon Oxide Films on Silicon. ELECTROCHEMICAL SOC., J., v. 115, no. 3, Mar. 1968. p. 300-307. [2]

DILL, H.G. and T.N. TOOMBS. A New MNOS Charge Storage Effect. SOLID STATE ELECTRONICS, v. 12, no. 12, Dec. 1969. p. 981-987. [1]

DILL, H.G. et al. Anomalous Behavior in Stacked-Gate MOS Tetrodes. IEEE INTERNATIONAL SOLID-STATE CIRCUITS CONF., 1969, Philadelphia, Penna. Paper WPM-4.4, p. 44-45. [2]

DONOVAN, R.P. Integrated Silicon Device Technology. v. 13, Intraconnections and Isolation. Research Triangle Institute, Report no. ASD-TDR-63-316, v. 13, May 1967. 155 p. AD 655 081.

DOO, V.Y. Silicon Nitride, A New Diffusion Mask. IEEE TRANS. ON ELECTRON DEVICES, v. ED 13, no. 7, July 1966. p. 561-563. [1]

DOO, V.Y. and D.R. KERR. Investigation of Refractory Dielectrics for Integrated Circuits. IBM Corp. Report no. NASA-CR-995, Mar. 1968. 73 p. [2]

DOO, V.Y. et al. Preparation and Properties of Pyrolytic Silicon Nitride. ELECTROCHEMICAL SOC., J., v. 113, no. 12, Dec. 1966. p. 1279-1281. [3]

DOO, V.Y. et al. Property Changes in Pyrolytic Silicon Nitride with Reactant Composition Changes. ELECTROCHEMICAL SOC., J. v. 115, no. 1, Jan. 1968. p. 61-64. [4]

DUFFY, M. and W. KERN. Preparation, Properties and Applications of Chemically Vapor Deposited Silicon Nitride Films RCA REVIEW, v. 31, no. 4, Dec. 1970, p. 742-753.

ELLIOTT, E. et al. Some Properties of Electron Beam Evaporated Si_3N_4 Films. THIN SOLID FILMS, v. 3, no. 6, June 1969. p. R47-R48.

FAIRCHILD SEMICONDUCTOR CO. Passing Marks. ELECTRONICS, v. 42, no. 14, July 7, 1969. p. 46-48.

110

FLAD, F.W. et al. The Application of MNOS Transistors in a Preset Counter with Nonvolatile Memory. IEEE INTERNATIONAL SOLID-STATE CIRCUITS CONF., 1969, Philadelphia, Penna. Paper WPM-4.5, p. 46-47.

FOSTER, J.E. and J.M. SWARTZ. Electrical Characteristics of the Silicon Nitride-Gallium Arsenide Interface. ELECTROCHEMICAL SOC., J., v. 117, no. 11, Nov. 1970 p. 1410-1417.

FRANK, R.I. and W.L. MOBERG. Reactively Sputtered Oxynitride as a Dielectric Material for Metal-Insulator-Metal Capacitors. ELECTROCHEMICAL SOC., J., v. 117, no. 4, Apr. 1970. p. 524-529.

FRÄNZ, I. and W. LANGHEINRICH. Distribution of Sodium in Silicon Nitride. SOLID STATE ELECTRONICS, v. 12, no. 3, Mar. 1969. p. 145-150.   [1]

FRÄNZ, I. and W. LANGHEINRICH. Silicon Nitride as a Mask in Indium Diffusion. SOLID STATE ELECTRONICS, v. 12, no. 1, Jan. 1969. p. 63-64.   [2]

FRÄNZ, I. and W. LANGHEINRICH. Silicon Nitride as a Mask in Phosphorus Diffusion. SOLID STATE ELECTRONICS, v. 12, no. 12, Dec. 1969. p. 955-962. [3]

FROHMAN-BENTCHKOWSKY, D. and M. LENZLINGER. Charge Transport and Storage in Metal-Nitride-Oxide-Silicon (MNOS) Structures. J. OF APPLIED PHYS., v. 40, no. 8, July 1969. p. 3307-3319.   [1]

FROHMAN-BENTCHKOWSKY, D. The Metal-Nitride-Oxide-Silicon (MNOS) Transistor-Characteristics and Applications. IEEE PROCEEDINGS, v. 58, no. 8, Aug. 1970. p. 1207-1219.   [2]

FROHMAN-BENTCHKOWSKY, D. and D.D. FORSYTHE. Reliability of MNOS Integrated Circuits. INTERNATIONAL ELECTRON DEVICES MEETING, Washington, D.C., Oct. 1969. p. 48.   [3]

GENERAL INSTRUMENT CORP. Reliability Aspects of Low Voltage Nitride. Company Report. Hicksville, N.Y. Aug. 1969. 11 p.

GOETZBERGER, A. and E.H. NICOLLIAN. Transient Voltage Breakdown Due to Avalanche in MIS Capacitors. APPLIED PHYSICS LETTERS, v. 9, no. 12, Dec. 1966. p. 444-446.

GOODMAN, A.M. et al. Optimization of Charge Storage in the MNOS Memory Device. RCA REVIEW, v. 31, no. 2, June 1970. p. 342-354.

GORDON, F. Jr. and H.E. WANNEMACHER Jr. The Effects of Space Radiation on MOSFET Devices and Some Application Implications of Those Effects. IEEE TRANS. ON NUCLEAR SCIENCE, v. NS-13, no. 6, Dec. 1966. p. 262-272.

GRAY, C. Silicon Nitride Shields for Plastic Encapsulated Transistors. 7th RELIABILITY PHYSICS SYMPOSIUM, 1968. p. 37. (Abstract)

GREGOR, L.V. Study of Silicon Nitride as a Dielectric Material for Micro-electronic Applications. IBM Corp., Final Report no. AFAL-TR-67-248. Sept. 1967. AD 824 221.   [1]

GREGOR, L.V. Study of Silicon Nitride as a Dielectric Material for Micro-electronic Applications. IBM Corp., Interim Report AF33(615)-5386-1. Nov. 1966. 36 p.   [2]

GREGOR, L.V. Silicon Nitride, Deposition and Application. THIN FILM DIELECTRICS, Ed. F. Vratny. Electrochemical Society, Dielectrics and Insulation Division, 1969. p. 447-488.   [3]

GRUBER, G.A. and R.R. VERDERBER. The Effects of Film Thickness and Deposition Temperature on SiO_2/Si_3N_4 Passivated High Voltage Mesa Diodes. THE ELECTRO-CHEMICAL SOCIETY, Los Angeles, May 10-15, 1970. Abstr. no. 115, p. 296-298.

GYULAI, J. et al. Outdiffusion through Silicon Oxide and Silicon Nitride Layers on Gallium Arsenide. APPLIED PHYSICS LETTERS, v. 17, no. 8, Oct. 1970. p. 332-334.

HARTMAN, D.K. and E.A. HERR. Silicon Nitride Manufacturing Process. General Electric Co., Syracuse, N.Y., Semiconductor Products Dept. Report No. IR-509-8-(1), Oct. 1968. 89 p. AD 845 764.   [1]

HARTMAN, D.K. and E.A. HERR. Silicon Nitride Manufacturing Process. General Electric Co., Syracuse, N.Y., Semiconductor Products Dept. Report No. IR-509-8-(2), Jan. 1969. 73 p. AD 846-954.   [2]

HEUMANN, F.K. et al. Diffusion Masking of Silicon Nitride and Silicon Oxynitride Films on Si. ELECTROCHEMICAL SOC., J., v. 115, no. 1, Jan. 1968. p. 99-101.

HILLERY, R.V. Changes in the Surface Charge Density of Thermally Oxidized Silicon due to Silicon Nitride Deposition. SEMICONDUCTOR SILICON. Ed: R.R. Haberecht and E.L. Kern. 1969. Electrochemical Society. p. 339-349. [1]

HILLERY, R.V. and R.L. CLARK. Effect of Process Variations on Masking Properties of Pyrolytically Deposited Silicon Nitride Films. THE ELECTRO-CHEMICAL SOCIETY, Los Angeles, May 10-15, 1970. p. 290-291.   [2]

HOUSE, S.S. and R.A. WHITNER. Manufacturing Beam Lead Sealed Junction Monolithic Integrated Circuits. WESTERN ELECTRIC ENGINEER. v. 11, no. 1, Jan. 1967. p. 3-15.

HSIA, Y. and G.S. HOLLAND. Silicon Nitride Non-Volatile Semiconductor Storage. WESTERN ELECTRON SHOW AND CONVENTION, (WESCON), LSI MEMORIES SESSION, Los Angeles, Aug. 26, 1970. 18 p.

HU, S.M. Properties of Amorphous Silicon Nitride Films. ELECTROCHEMICAL SOC., J., v. 113, no. 7, July 1966. p. 693-698.   [1]

HU, S.M. et al. Evidence of Hole Injection and Trapping in Silicon Nitride Films Prepared by Reactive Sputtering. APPLIED PHYSICS LETTERS, v. 10, no. 3, Feb. 1967. p. 97-99. [2]

HU, S.M. and L.V. GREGOR. Silicon Nitride Films by Reactive Sputtering. ELECTROCHEMICAL SOC., J., v. 114, no. 8, Aug. 1967. p. 826-833. [3]

HUTCHINS, C.L. Charge Transients in Aluminum-Silicon Nitride-Silicon Capacitors. Ph.D. Thesis. North Carolina State College, Raleigh, N.C., Semiconductor Device Lab. Feb. 1968. 176 p. X68-16716. [1]

HUTCHINS, C.L. and R.W. LADE. Charge Storage in Metal-Silicon Nitride-Silicon Capacitors. IEEE PROC., v. 55, no. 8, Aug. 1967. p. 1494-1495. [2]

IBM CORP., A Dielectric Material for Study of Silicon Nitride in Microelectronic Applications. Nov. 1968. AD 843 876.

JONES, R.E. and V.Y. DOO. Integrated Circuit Isolation with Silicon Nitride. ELECTROCHEMICAL TECHNOLOGY, v. 5, May/June 1967. p. 297-298.

KENDALL, E.J.M. The Stabilization of Silicon Surfaces Using Silicon Nitride. J. OF PHYSICS, PART D, BRITISH J. OF APPLIED PHYSICS, series 2, v. 1, no. 11, Nov. 1968. p. 1409-1420.

KESHAVAN, B. et al. Large Scale Integration of MNOS Devices. INTERNATIONAL ELECTRON DEVICES MEETING, IEEE ELECTRON DEVICES GROUP, Oct. 1967. p. 18-19. (Abstract) [1]

KESHAVAN, B. and H.C. LIN. MONOS Memory Element. INTERNATIONAL ELECTRON DEVICES MEETING, IEEE ELECTRON DEVICES GROUP, Oct. 1968. p. 141-142. [2]

LAWRENCE, H. and P.C. SCHAEFER. A Three Masking Step MIS Process. ELECTRO-CHEMICAL SOCIETY, Los Angeles Calif., May 10-15, 1970. p. 359-361.

LEGAT, W.H. Application of Sputtering in the Fabrication of Semiconductor Devices. TRANSACTIONS ON SPUTTERING ELECTRONICS, Jan. 1970. Pebble Beach, Calif.

LEPSELTER, M.P. Beam-Lead Sealed-Junction Technology. BELL LABORATORIES RECORD, v. 44, no. 9, Oct./Nov. 1966. p. 298-303.

LEWIS, E. et al. Investigation of New Concepts of Adaptive Devices. SPERRY RAND RES. CTRE., 7th Quarterly Tech. Rept., SRRC-69-16, NASA-CR-86255, June 1969. 16 p. NASA-N69-40438. [1]

LEWIS, E. et al. Investigation of New Concepts of Adaptive Devices. SPERRY RAND RES. CTRE., Rept. no. SRRC-CR-68-20, NASA-CR-86251, Apr. 1968, 15 p. NASA-N69-40361. [2]

LEWIS, E. and H.A.R. WEGENER. Investigation of New Concepts of Adaptive Devices. SPERRY RAND RES. CTRE., 5th Quarterly Tech. Rept., NASA-CR-86253, SRRC-CR-68-51, Dec. 1968. 10 p. NASA N69-40271. [3]

LEWIS, E. et al. Investigation of New Concepts of Adaptive Devices. SPERRY RAND RES. CTRE., Rept. no. SRRC-CR-69-7, NASA-CR-86254. Apr. 1969. 17 p. NASA-N69-40066. [4]

McDONALD, B.A. The MNOS Bipolar Transistor. INTERNATIONAL ELECTRON DEVICE MEETING, IEEE ELECTRON DEVICES GROUP, Washington, D.C. Oct. 1970.

MILEK, J.T. Silicon Nitride for Microelectronic Applications. Part I. Preparation and Properties. HANDBOOK OF ELECTRONIC MATERIALS, v. 3, N.Y. Plenum Press, 1971. 118 p.

MOTOROLA CO. Nitride-Passivated Transistors Marketed. CHEMICAL ENGINEERING NEWS, v. 46, no. 30. July 15, 1968. p. 54-55.

MYERS, T.R. Silicon Nitride Surface Passivation. RELIABILITY ANALYSIS CENTER. Technical Monograph 69-2, Sept. 1969. 36 p.

NABER, C.T. A Method for Reducing the Mobile Electric Charge in MNOS Structures. ELECTROCHEMICAL SOC., J., v. 116, no. 9, Sept. 1969. p. 1282-1284.

NEUGEBAUER, C.A. et al. Polycrystalline CdS Thin Film Field Effect Transistors: Fabrication, Stability and Temperature Dependence. THIN SOLID FILMS, v. 2, no. 1, Jan. 1968. p. 57-78.

NEWMAN, P.A. and H.A.R. WEGENER. Effect of Electron Radiation on Silicon Nitride Insulated Gate Field Effect Transistors. IEEE TRANS. ON NUCLEAR SCIENCE, v. NS-14, no. 6, Dec. 1967. p. 293-298.

PAO, H.C. and M. O'CONNELL. Memory Behavior of an MNS Capacitor. APPLIED PHYSICS LETTERS, v. 12, no. 8, April 1968. p. 260-263.

PATTERSON, W.L. Memories--Digging in Data. ELECTRONICS, v. 42, no. 8, April 14, 1969. p. 50-52.

PERKINS, C.W. et al. Radiation Effects and Electrical Stability of Metal-Nitride-Oxide-Silicon Structures. APPLIED PHYSICS LETTERS, v. 12, no. 11, June 1968. p. 385-387. [1]

PERKINS, C.W. et al. Radiation Effects in Modified Oxide Insulators in MOS Structures. IEEE TRANSACTIONS ON NUCLEAR SCIENCE, v. NS-15, no. 6, Dec. 1968. p. 176-181. [2]

PLESSEY LTD. Plessey Develops MNOS Read-Mostly Memory. ELECTRONICS, v. 43, no. 25. Dec. 7, 1970. p. 159.

RABURN, W.D. An Investigation of Metal Insulator Interfaces. Alabama Univ., University, Ala., Bureau of Engineering Research. Final Rept., June 1970. 31 p. N70-35950.

ROSS, E.C. et al. Optimization for MNOS Device Performance. INTERNATIONAL ELECTRON DEVICES MEETING, IEEE ELECTRON DEVICES GROUP. Oct. 1969. p. 46. (Abstract) [1]

ROSS, E.C. et al. Effects of Silicon Nitride Growth Temperature on Charge Storage in the MNOS Structure. APPLIED PHYSICS LETTERS, v. 15, no. 12, Dec. 1969. p. 408-409. [2]

ROSS, E.C. and J.T. WALLMARK. Theory of the Switching Behavior of MIS Memory Transistors. RCA REVIEW, v. 30, no. 2, June 1969. p. 366-381. [3]

ROSS, E.C. et al. Operational Dependence of the Direct-Tunneling Mode MNOS Memory Transistor on the SiO_2 Layer Thickness. RCA REVIEW, v. 31, no. 3, Sept. 1970. p. 467-478. [4]

SARACE, J.C. et al. Metal-Nitride-Oxide-Silicon Field-Effect Transistors, with Self-Aligned Gates. SOLID STATE ELECTRONICS, v. 11, no. 7, July 1968. p. 653-660.

SAVAGE, J.A. et al. Investigation of Silicon Nitride Thin Films for Use in MOST Devices. ROYAL RADAR ESTABLISHMENT NEWSLETTER AND RESEARCH REVIEW, Malvern, England. No. 7. 1968. p. 15/1-2. N69-28679.

SCHMIDT, P.F. et al. Radiation Insensitive Silicon Oxynitride Films for Use in Silicon Devices. IEEE TRANSACTIONS ON NUCLEAR SCIENCE, v NS-16, no. 6, June 1969. p. 211-219. [1]

SCHMIDT, P.F. and J.D. ASHNER. Radiation Insensitive Silicon Oxynitride Films for Use in Silicon Devices. Part II. IEEE TRANSACTIONS ON NUCLEAR SCIENCE, v. NS-17, no. 6, Dec. 1970. p. 11-17. [2]

SCHMIDT, P.F. and D.R. WONSIDLER. Conversion of Silicon Nitride Films to Anodic SiO_2. ELECTROCHEMICAL SOC., J., v. 114, no. 6, June 1967. p. 603-605. [3]

SCHNEER, G.H. et al. A metal-Insulator-Silicon Junction Seal. IEEE TRANSACTIONS ON ELECTRON DEVICES, v. ED-15, no. 5, May 1968. p. 290-293.

SEDGWICK, T.O. et al. Dielectric Films for Ge Planar Devices. IBM JOURNAL OF RESEARCH AND DEVELOPMENT, v. 14, no. 1, Jan. 1970. p. 2-11.

SEWELL, F.A. Jr. An MNS Light Sensitive Memory Element. INTERNATIONAL ELECTRON DEVICES MEETING, IEEE ELECTRON DEVICES GROUP, Washington, D.C. Oct. 1969. p. 46-48. (Abstract) [1]

SEWELL, F.A. Jr. et al. Charge Storage Model for Variable Threshold FET Memory Element. APPLIED PHYSICS LETTERS, v. 14, no. 2, Jan. 1969. p. 45-47. [2]

SEWELL, F.A. Jr. et al. The Variable Threshold FET-Theory and Experiment. IEEE INTERNATIONAL SOLID STATE CIRCUITS CONFERENCE, Philadelphia, Penna., 1969. p. 182-183. [3]

SEWELL, F.A. Jr. et al. Investigation of New Concepts of Adaptive Devices. SPERRY RAND RESEARCH CENTRE, Rept. no. SRRC-CR-68-8, 2nd Quarterly Rept., Jan. 1968. 13 p. N69-40201. [4]

SEWELL, F.A. Jr. et al. Metal-Insulator-Semiconductor Transistor for Use as a Nonvolatile Digital Storage Element. SPERRY RAND RESEARCH CENTRE, Technical Rept., AFAL-TR-70-148, 86 p. AD 872-991. [5]

SOREF, R.A. High-gain, Low-noise Infrared Photodetectors. SPERRY RAND RESEARCH CENTRE, Final Rept. no. SRRC-CR-68-25, May 1968. 49 p. AD 832-963.

STAFFORD, K.R. and D.W. OBERLIN. A Monolithic Radiation-hardened Operational Amplifier. SOLID STATE TECHNOLOGY, v. 13, no. 5, May 1970. p. 67-72.

STANDARD TELECOMMUNICATION LABORATORIES. Interconnecting via Multilayers. ELECTROTECHNOLOGY, v. 84, no. 5, Nov. 1969. p. 16-17.

STANLEY, A.G. Comparison of MOS and Metal-Nitride-Semiconductor Insulated-Gate Field Effect Transistors under Electron Irradiation. IEEE TRANSACTIONS ON NUCLEAR SCIENCE, v. NS-13, no. 6, Dec. 1966. p. 248-254.

STOLLER, A.I. et al. A Novel Technique for Forming Glass-to-Metal Seals Using a Silicon Nitride Interface Layer. RCA REVIEW, v. 31, no. 2, June 1970. p. 443-449.

SUGUWARA, K. Facets Formed by HCl Vapor Etching on Silicon Surfaces through Windows in SiO_2 and Si_3N_4 Masks. ELECTROCHEMICAL SOC., J., v. 118, no. 1, Jan. 1971. p. 110-114.

SWANN, R.C.G. et al. The Preparation and Properties of Thin Film Silicon-Nitrogen Compounds Produced by a Radio Frequency Glow Discharge Reaction. ELECTROCHEMICAL SOC., J., v. 114, no. 7, July 1967. p. 713-717.

TOOMBS, N.C. et al. A New Insulated-Gate Silicon Transistor. IEEE PROCEEDINGS, v. 54, no. 1, Jan. 1966. p. 87-89.

TOTAH, R.P. Effect of Radiation on the Stability of Silicon Nitride and Metal Nitride Semiconductor FET Devices. HIGHES AIRCRAFT CO., Final Report, NAS 8-18003, April 1967. 77p.

TRAPP, G.D. and J.B. PREECE. Silicon Nitride Passivated Integrated Circuits-Reliability Improvements. 6th Annual Reliability Physics Symposium, Los Angeles, Calif., Nov. 1967. p. 96-105.

VADASZ, L.L. et al. Silicon-Gate Technology. IEEE SPECTRUM, v. 6, no. 10, Oct. 1969. p. 28-35.

VERDERBER, R.R. et al. SiO_2/Si_3N_4 Passivation of High-Power Rectifiers. IEEE TRANSACTIONS ON ELECTRON DEVICES, v. ED-17, no. 9, Sept. 1970. p. 797-799.

WALLMARK, J.T. and J.H. SCOTT. Switching and Storage Characteristics of MIS Memory Transistors. RCA REVIEW, v. 30, no. 2, June 1969. p. 335-365.

WEGENER, H.A.R. Investigation of New Concepts of Adaptive Devices. SPERRY
RAND RESEARCH CENTRE, Report no. SRRC-CR-68-43, Sept. 1968, 122p. N69-14097.
[1]

WEGENER, H.A.R. MNOS Memories. IEEE INTERMAGNETIC CONFERENCE, Washington,
D.C., April 1970. Paper no. 11.5 in 1970 Digest of Conference. 2 p.   [2]

WEGENER, H.A.R. et al. Investigation of New Concepts of Adaptive Devices.
SPERRY RAND RESEARCH CENTRE, Report No. SRRC-67-62, Dec. 1967. 30 p.
N69-40402.   [3]

WEGENER, H.A.R. and F.A. SEWELL Jr. Metal-Insulator-Semiconductor Transistor
for Use as a Nonvolatile Digital Storage Element. SPERRY RAND RESEARCH CENTRE,
Tech. Rept. AFAL-TR-69-187, July 1969. 78 p. AD 856 260.   [4]

WEISS, J.S. Inexpensive Electronic Memories. PROCEEDINGS OF THE SID, v. 11,
no. 3, 3rd Quarter 1970. p. 97-104.

WESTINGHOUSE MOLECULAR ELECTRONICS DIVISION, Elkridge, Md. Increased
Reliability for Plastic-Packaged ICs. ELECTRONIC PRODUCTS, v. 10, no. 12,
May 1968. 85 p.

ZAININGER, K.H. Irradiation of MIS Capacitors with High Energy Electrons.
IEEE TRANSACTIONS ON NUCLEAR SCIENCE, v. NS-13, no. 6, Dec. 1966. p. 237-246.

ZIMBELMANN, H.P. and L.S. PENNER. Custom Design with MTNS. NEREM RECORD,
1969. p. 126.